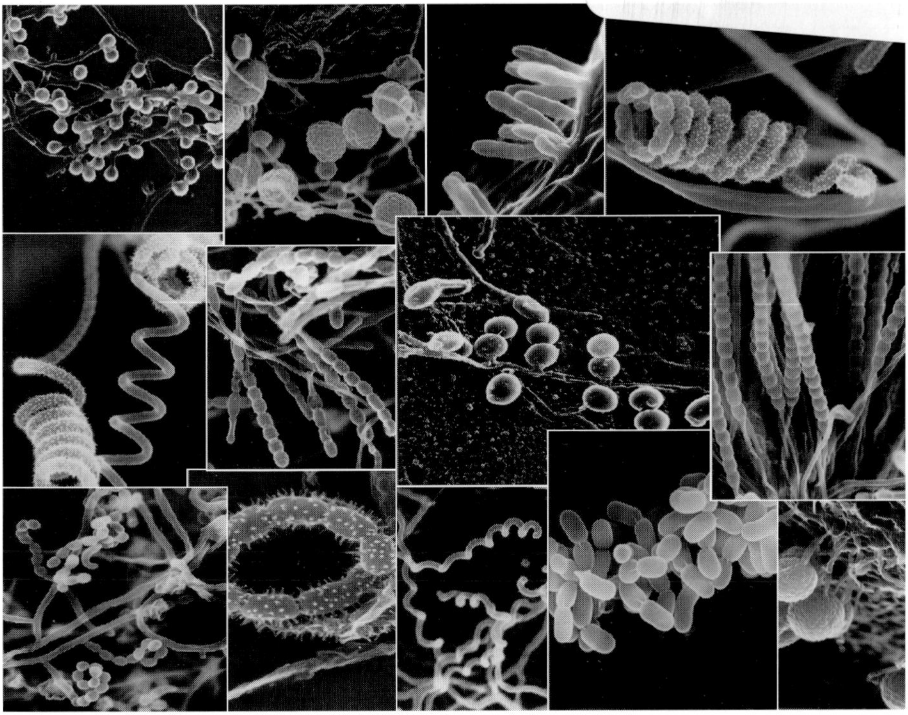

Microbial Genetic Resources and Biodiscovery

Edited by

Ipek Kurtböke and Jean Swings

Microbial Genetic Resources and Biodiscovery

Edited by

Ipek Kurtböke and Jean Swings

First printed 2004

National Library of Australia

ISBN 0 646 44048 9

Printed by

Queensland Complete Printing Services
28 Price Street, Nambour, Queensland
www.qprint.com.au

Cover Photo: Bioactive compound producing Actinomycetes

Courtesy of Shinji Miyadoh, Japan

A World Federation of Culture Collections (WFCC)

Publication

In Collaboration with

The Institute for Sustainability, Health and Regional Engagement, University of the Sunshine Coast, Australia

Sale enquiries should be directed to

Dr David Smith

WFCC Secretariat
c/o CABI Bioscience UK Centre (Egham)
Bakeham Lane
Egham
Surrey TW20 9TY
UK

Tel: +44-1491 829080
Fax: +44-1491 829100
Email: d.smith@cabi.org

Contents

Philosophy of new drug discovery relies on the "(i) belief in the great capabilities of the microorganisms, (ii) development of well-designed screening systems against given objectives, (iii) recognition that screening is not a mere routine operation, (iv) emphasis on basic research and (v) treasuring good human relationships".

Omura (1986)

Omura, S. (1986). Philosophy of new drug discovery. *Microbiological Reviews,* **50:** 259-279.

Microbial Genetic Resources, Biodiscovery and Intellectual Property Rights

Microbial Genetic Resources and Biodiscovery

D.I. Kurtböke[1], J. Swings[2] and V. Storms[2]

[1]: *Institute for Sustainability, Health and Regional Engagement, University of the Sunshine Coast, Maroochydore DC, QLD 4558, Australia*

[2]: *BCCM/LMG Bacteria Collection and Laboratory for Microbiology, Faculty of Science, University of Gent, Belgium*

1. Microbial Genetic Resources

1.1. Microbial diversity in the context of biodiscovery and biotechnology

Microorganisms are the most numerous and ancient biotic entities which have successfully colonized every possible ecological niche on the planet. Their presence and activity is essential to the healthy functioning of ecosystems [3]. The role they play in nature to maintain the dynamic equilibrium and integrity of the biosphere is of such magnitude that the continued existence of biological life itself is crucially dependent on the sustained microbe-mediated transformation of matter in terrestrial and aquatic ecosystems [3]. The extraordinary activity of microorganisms is based on their remarkable metabolic diversity and genetic adaptability [3, 15] and as a result the interest subsequently widened in the development of microbial technologies for the production of antibiotics and other therapeutic agents, management of pests and pathogens, bioleaching of metals, increasing soil fertility, generating biofuels, monitoring air pollution, destroying persistent pollutants, waste water treatment, bioremediation, and serving tools in biomedical research [3].

The use of microbially derived compounds has now been extended to other areas of industrial applications such as the commercial production of sugars syrups from agricultural products representing the principal commercial function of enzymes [24, 26], where they have a large impact in reducing environmental pollution. Currently, the major markets are detergent enzymes as well as food and feed processing. However, pulp, paper, textile and leather sectors are the fastest growing markets for the microbial derived industrial enzymes [21, 34].

As a result, the exploitation of microorganisms in industry is worth billions of dollars each year [40]. Therefore, it is believed that the possibility to increase the chemical diversity of the compounds integrated into biotechnological programs for different industrial sectors exists through the exploitation of microbial diversity [7].

Over the past three decades, biotechnology has benefited greatly from the extraordinary advances in molecular biology but the previously advocated counterbalance available from biodiversity has strengthened the force of the argument during the intervening period [5, 6]. As stated by Bull [5, 6] the unique opportunities provided by recombinant technologies are indisputable, but for many biotechnological products and processes this option is unnecessary, inappropriate, or even proscribed. The demand for totally novel or replacement products, on the other hand, exemplifies the need for biotechnologies based upon natural organisms, in particular those microorganisms with defined/targeted activity [5, 6].

1.2. Search for novel microorganisms of biotechnological value

1.2.1. The role of taxonomy

The role of taxonomic methods in the search and discovery of new bioactive compounds is increasingly becoming appreciated by industry, as also is the need for greater numbers of trained microbial systematics [41]. The case for investing in microbial systematics as an effective route to isolation, screening, and the recognition of novel taxa has been continuously emphasized [30, 41]. In the past, taxonomic studies conducted on the strains shown to be producing new substances illustrated that they were close relatives of the taxa and similarly yielded novel exploitable properties. Examples are the discovery of the broad specificity herbicide (phosalacine), and an inhibitor of cell wall synthesis (setomimycin) from the members of the genus *Kitasatosporia* [5, 29].

In addition, the antibiotic biosynthetic gene transfer in streptomycetes has been demonstrated [8], suggesting that lateral gene transfer plays a significant role in the distribution of natural product biosynthesis across biodiversity [41]. The metabolic profiles of the well-characterized members of the *Streptomyces violaceusniger* clade were determined and they all showed the same metabolic pattern of HPLC-detected secondary metabolites consisting of geldanamycin, eliaophylin, nigericin and a characteristic polyene, a set of metabolites consistent with those previously found in *Streptomyces hygroscopicus* strains [1, 9, 41].

Watve *et al.*, [42] using mathematical modeling studies have demonstrated that an estimate of over 150,000 bioactive metabolites are still waiting to be discovered from the members of the Genus *Streptomyces*. Consequently, systematic studies defining the antibiotic producing clusters of this taxon will facilitate the rapid detection of the bioactive compound-producing novel *Streptomyces* spp. [41].

Ward and Goodfellow [41] have currently stated that systematics, including taxonomy, like the rest of biology, is currently in the throes of a technological revolution. Developments in the high throughput sequencing are providing unprecedented amounts of data. High throughput techniques for phenotypic characterization, whole genome and multi-locus sequencing, micro-arrays and subtractive hybridization will define the unit of diversity for prokaryotes and will provide rapid and accurate methods for dereplication and identification. They will also help build up a roadmap for search and biodiscovery strategies to detect and identify organisms and define ecological niches in terms of the organisms present.

1.2.2. The role of molecular biology

The role of molecular biology techniques as a means of enquiry into the nature and into the extent of microbial diversity has been significant. They have been used successfully to reveal microbial diversity with significant novelty [44]. Using taxon specific probes, scientists now easily detect novel members of the taxon of interest even in well-studied ecosystems. The implications of such findings are profound for ecologists, and also for biotechnologists who will want to screen the newly revealed genetic diversity for useful properties [5, 6].

1.2.3. The role of selective isolation

Molecular ecology has undoubtedly revealed the extent of our ignorance but it has still not provided that the ability to detect and identify organisms and define ecological niches in organismal terms [41]. Therefore, once the microbial genetic diversity is revealed through molecular microbial ecological techniques, using the information gained from the cultured members of the taxon under investigation, target directed selective isolation techniques could be designed. Many novel actinomycete taxa with novel therapeutic activities have been cultured using such highly selective isolation techniques [22, 43].

The use of genus-oriented isolation methods in industrial programs has provided significant impetus towards the discovery of new bioactive compounds [23]. Moreover, these methods have facilitated the progress in the fields of actinomycete ecology, taxonomy and phylogeny [22].

Accordingly, the availability of high fidelity microbial strain databases from the cultured members of the microbial taxa of interest, offer real opportunities for the design of media for selective isolation and the detection of novel or rare microroganisms. The availability of computer programs also provides a rational basis for formulating taxon-specific isolation media [43]. As stated previously [22] selective isolation of biotechnologically important microorganisms is no longer the application of a single technique but it is a design of methodologies using the information generated through eco-taxonomical disciplines.

1.2.4. Conservation of the microbial genetic pool

Microorganisms are now becoming increasingly significant in national economies; however, still a vast number of microbial species are yet to be identified [22, 41]. It is ironic that, despite the recognition of their commercial value so little is known about the extent of their diversity and the complexity of their functional roles in sustaining global "life-support systems" such as agricultural, forest, coastal and freshwater ecosystems. Thus, as indicated by Bull [5, 6] a *prima facie* case for conservation can be upheld on the grounds of preserving and utilizing microbial diversity.

In principle, every genotype and every constituent DNA sequence is unique, irreplaceable and hence deserves conservation [27]. However, microbial genetic diversity is so broad and dynamic that an attempt to capture it in its entirety would be unrealistic. Also, the financial resources required would make it impossible to preserve all existing microbial resources. For that reason, we need to establish the priorities.

Culture collections are known to be the cornerstone of biotechnology. However, reliable and adequate collections are needed if biotechnological processes and innovations are to be furthered. Culture collections play an important role in the exchange of information on cultures as a significant step towards sustainable development. Culture collections must be kept aware of current developments in microbiology and biotechnology so that they can focus on maintaining and providing germplasm, the data most needed [15, 16, 27, 35, 36].

Culture collections are mainly built up through isolations carried out in natural habitats, supplemented by type or other reference strains acquired from other collections. There are unfortunately no data on how effectively various natural habitats have been eco-geographically surveyed for microorganisms. Natural habitats frequently contain microorganisms with useful traits that would have disappeared if they had not been collected and safely preserved in an unchanged form. While their occurrence in vast numbers, rapid reproduction, and broad genetic diversity do not allow specific microbial traits to become extinct unless they are subjected to radical environmental changes as a result of pollution, concentrated accumulation of organic wastes, massive application of chemical fertilizers and pesticides, destruction of ecosystems, and so on. However, in practice, retrieval of particular microbial strains exhibiting specific traits from natural ecosystems is almost impossible. Undoubtedly, safe conservation of microbial strains with desired traits in service culture collections is the only way of making such a trait readily available for exploitation. But research is necessary to ensure that strains with valuable traits may not be lost during isolation, preservation, or storage. Moreover, the loss of microbial habitats due to constantly growing human settlement makes it impossible to reisolate and replace lost material. Accordingly, it is useful to bear in mind that geographic, environmental, geological and climatic factors lead to a significant differentiation of microbial flora from one location to another ensuring the versatility of microbial strains. Intact habitats in tropical and subtropical zones constitute the areas of greatest microbial genetic diversity however these are often found in economically disadvantaged countries lacking the necessary resources to initiate programs for the isolation, characterization, preservation, and exploitation of the available material [27].

While the threat to biodiversity is being documented with increasing regularity and purpose today, it still continues. This threat has its origins in environmental mismanagement resulting from chemical pollution, overexploitation of natural renewable and non-renewable resources, large scale physical interventions leading to changes in land and water use and population growth. The scale of the threat has to be considered in relation to animal and plant species, and to habitats and tropical forests, wetlands, and coral reefs, are ecosystems of extreme diversity and consequently excite particular concern. Consequences of the loss of genetic diversity in animals and plants are very significant for biotechnology. Bull [5] has illustrated that in broad terms only about 1.5% (4000) of flowering plants has been screened for pharmaceutical compounds, and of this number about 3% currently provide major drugs. The undiscovered potential is thus highly considerable. The rate of plant species loss due to tropical deforestation is estimated at 0.6% per annum, with about 25% of plant species likely to be lost within 20 years. Prediction of the loss of microbial diversity is much more difficult due to scanty information and anecdotal evidence for loss. Yet, loss occurs. For example, research has shown that *Penicilliopsis claviaeformis* disappeared from the Bogor Botanic Gardens as the appropriate *Diospyros* tree species with which it is associated was no longer available. Similarly, the virtual disappearance of *Cookenia tricholoma* from Java was a consequence of interference with the island's ecosystem. Many microorganisms appear resistant to environmental changes. However, those found in unspoiled settings and closely associated with microorganisms may be less ubiquitous, and thereby susceptible to habitat disturbance [5].

Traditionally microorganisms have been maintained in culture collections or herbaria. However, *in situ* conservation is as crucial for the protection of microbial genetic diversity as it is for the protection of macroorganisms. Thus the message: "conserving biological diversity equals conserving ecosystems" [5, 19, 20] has the same validity for microbiologists and biotechnologists as it does for botanists and zoologists. **The World Conservation Strategy** [5, 18] was only partially correct when it said "isolating from nature a microbial strain with a particular desired property is tedious" and "such strains need to be maintained in pure culture collections". It is important to emphasize that culture collections fulfill several vital requirements for biotechnology [5, 6] including collection-oriented research and collection-derived new products [33].

1.3. Searching for novel biotechnological properties

The search for new pharmaceuticals and other biologically useful products has led to bioprospecting for high value drugs and enzymes. Traditionally, biodiscovery of natural products has been a lengthy process involving a number of stages. Firstly, finding novel microorganisms for testing requires extensive search of previously unexplored habitats. Then, researchers must create extracts that can be screened for a desired activity, such as inhibiting cell growth for a candidate cancer drug. Next, the active compound has to be isolated, purified, and tested again. Finally, if the compound passes all these stages, researchers must hope that it has not already been identified and patented by other research teams [10, 14].

In order to shorten the traditionally lengthy process, pharmaceutical companies have developed alternative approaches to biodiscovery. For example, Novo-Nordisk in Denmark has developed a prescreening system to ensure the chemical diversity of the extract collection. They established a HPLC library containing different classes of reference compounds such as macrolides, peptides and quinones. A neural network was also developed for proper evaluation of the high amount of HPLC data. Using this network it was possible to reduce the HPLC library search time for each extract more than 10 times. Furthermore, they were able to identify media dependent compounds. Such technology enabled the screening of higher number of microbial extracts. It also ensured the investigation of the potential of different microorganisms to produce different types of metabolites, the selection of optimal fermentation and sample preparation conditions for new groups of microorganisms and the chemical diversity of the microbial extracts used in screening [14].

Since early 1990s, modern search and discovery programs have placed emphsis on the development of target-directed screens that produce high probabilities of success [5].

Approaches included:

- Mode of action screens
- Enzyme inhibitors
- Ligand mimicry
- Immunoregulation
- Roles for genetics and molecular biology in target-directed screens
- Automation
- The use of parameter regulation

Current discovery programs consider four key elements to conduct a successful drug discovery program [14]:

- number,
- diversity,
- novelty of the compounds entering the program,
- the speed by which it can be done

In addition, the focus is also on four different sources to be used in the screens, which are peptides, defined chemicals, combinatorial chemicals and natural products. Of the four sources for new drug discovery programs, the natural products have been proven to be valuable for new compounds originating from plants, marine macroorganisms and microorganisms.

Most pharmaceutical discovery programs tend to focus on microorganisms due to the following facts:

- They are known to be capable of producing industrially interesting compounds
- New microorganisms tend to produce new types of molecules
- Microorganisms possess unsurpassed diversity and the vast majority still remains to be discovered [42].

To optimize the chemical diversity, it is important not only to focus on the use of various microorganisms, but also to include the elements of various fermentation conditions and sample preparation techniques [14]. Therefore, the outcome of chemical and biological screening should be analyzed alongside ecological and taxonomical data on the microorganisms. This again emphasizes the importance of generating reliable databases on the eco-taxonomical characteristics of the microorganisms [22].

1.3.1. Combinatorial chemistry

Combinatorial chemistry synthesizes tens of thousands to millions of compounds at once. Robotic high-throughput techniques screen them en masse for drug leads. In this way, huge libraries of slightly differing compounds are compiled to help optimize drug lead.

Some pharmaceutical companies have thus turned away from 'empirical' discovery, which screened large numbers of chemical compounds for activity against therapeutic targets, to 'rational' discovery, that is the design of specific chemical compounds to act as molecular antagonists in disease processes believed to be much faster and cost effective.

The limitations of this rational discovery however, have sparked a renewed interest in natural product drug discovery. Combinatorial chemistry's promise to fill drug development pipelines with *de novo* synthetic small molecule drug candidates has not been fulfilled. Natural products have at least one major advantage over other chemical sources and strategies of lead findings. Specifically, the diverse chemical structures occurring naturally exceeds the scope of chemical synthesis and are not limited to areas of previous pharmacological activity as is almost always the case with synthetic libraries. Meanwhile, the practical difficulties of natural products drug discovery are being overcome by advances in separation technology and in the speed and sensitivity of structure elucidation.

1.3.2. Computer aided drug design and genomics

Recent advances in genomics have provided an opportunity to expand the range of potential drug targets and have facilitated a fundamental shift from direct screening programs towards rational target-based strategies. The application of genome-based technologies such as expression profiling and proteomics will lead to further changes in the drug discovery paradigm by combining the strengths and advantages of both screening strategies in a single program. Thus, the discovery of new drug targets through developments in genomics complements and extends the traditional genetic pool in providing a basis for rational drug-target discovery in anti-infectives [2, 17].

1.3.3. Unexplored biodiversity

Most pharmaceutical companies still remain committed to natural products biodiscovery and believe in the intrinsic utility of natural products as sources of drug leads [23, 37]. In addition, there is awareness that only a small fraction of the microbial diversity has been explored yet [23, 37].

For soil organisms, for instance, only 1% or less of the total population can be cultured by standard culturing techniques [22]. The number of insects described is about 1 million and many more have never been described. Most of them communicate or defend themselves through chemistry and only few have been studied by now. Most marine microorganisms probably have not been described, and those already described have not been examined chemically [6].

The plant endophyte niche, which remains mostly unexplored to date, is now revealing new bioactive compounds [38]. There are perhaps as many as 500,000 distinct varieties of higher plants and the variety of fungi associated with such plants is estimated to be at least as high, suggesting that the extent of this biodiversity is enormous [12]. Also the huge array of microbial diversity in extreme habitats, with enzymes functioning at extremes of temperature, salt, pH, pressure and other factors, serves as a resource waiting to be tapped and used for improved industrial processes and products [6, 11, 23, 29, 42].

Considering that so many microorganisms with biotechnological potential are waiting to be discovered in the natural ecosystems, it is obvious that the bioprospectors are still likely to find numerous new products of great economic value. However, this brings us to another important point in the drug discovery as the random bioprospecting has resulted in controversies about intellectual property rights and the control of access to biodiversity [4, 13, 31, 32, 39].

1.4. The Convention on Biological Diversity (CBD)

At the Earth Summit in Rio de Janeiro in 1992, world leaders agreed on a comprehensive strategy for "sustainable development". One of the key agreements adopted was the *Convention on Biological Diversity* (www.biodiv.org). The Convention establishes three main goals:

- the conservation of biological diversity,
- the sustainable use of its components,
- the fair and equitable sharing of the benefits from the use of genetic resources.

Thus the process of obtaining resources trough the use of biodiversity in commercial products involves negotiations and contracting with industrialized enterprises [39].

Costa Rica was the first country to implement this and believes that countries that are able to establish the most efficient regulations for the conservation of their biodiversity will be in better conditions to negotiate the acquisition of more benefits [39]. The National Institute of Biodiversity of Costa Rica (INBio) (http://www.inbio.ac.cr) is a leader in the negotiation and signature of benefit sharing agreements. Since its opening in 1989, thus even before the CBD, INBio has signed more than 10 agreements of this nature [39].

The best-known benefit sharing agreement is the 'INBio-Merck Agreement', the parties conceived the agreement as a contract to collaborate in the investigation of the existent biodiversity in Costa Rica's tropical forests in order to establish its potential applications to human health and animals. Bioprospecting agreements stipulate that 10% of research budgets and 50% of any future royalties be awarded to the Ministry of Environment and Energy for reinvestment in conservation. The remainder of the research budget supports in-country scientific and processing activities, also oriented to conservation and the sustainable use of biodiversity [39].

A recent UN study reporting on Antarctica states that a biotechnology 'gold-rush' is threatening an unregulated harvest of extremophile bacteria, fungi and algae (http://www.ias.unu.edu/binaries/UNUIAS_AntarcticaReport.pdf). The unfettered search for life at the ends of the Earth threatens to undermine the international rules on intellectual property rights and could pose a fundamental threat to the fragile environment of the Antarctic. Part of the problem is that nobody owns the genetic resources of the Antarctic; it is neither international territory nor is it clearly within national jurisdiction (http://www.commondreams.org/headlines04/0202-03.htm), (http://www.millennium-debate.org/ind2feb046.htm).

The conservation of each country's biodiversity can be achieved in various ways. "*In-situ*" conservation focuses on conserving genes, species, and ecosystems in their natural surroundings. "*Ex-situ*" conservation uses zoos, botanical gardens, culture collections and gene banks to conserve species, preferentially in the country of origin. For both *ex-situ* conservation, as well as the protection of the material rights, culture collections play a major role (http://wdcm.nig.ac.jp/wfcc/).

1.6. The World Federation for Culture Collections (WFCC)

The WFCC is the largest independent global organization that represents professional individuals and culture collections, which preserve biodiversity and enable their proper use. They target living microorganisms, cell lines, viruses and parts and derivatives of them [35] (http://www.wfcc.info/).

The task of maintaining representative samples of microbial diversity cannot be achieved by one collection alone and it is imperative that organisms utilized in biotechnology are maintained in a way that they retain their full capacity. It is essential that a worldwide network of collections such as WFCC interact to provide the coverage required by the user [35]. Besides, culture collections also play a capacity-building role, to help biodiversity-rich countries better understand and utilize their microbial diversity, and an important resource for public information and policy formulation. They are bodies that the public and the policy makers can call upon for objective help in developing regulations and guidelines for the safe and ethical use of biological resources (http://www.wfcc.info/).

The WFCC is linked to and participates actively in a large number of international organizations and initiatives e.g. the Organization for Economic Co-operation and Development (OECD; www.oecd.org) BRC initiative, the Global Biodiversity

Information Facility (GBIF; www.gbif.org), the World Intellectual Property Organization (WIPO, www.wipo.org) access and benefit sharing.

In addition, the WFCC together with the CBD is now looking into the development of a synergistic capacity to generate and enhance programs at the national, regional, and global level to increase the currently limited knowledge on the microbial diversity and in turn to strengthen the role of microbial culture collections as tools of research in biodiversity and biosystematics [16, 35].

1.7. From culture collections to Biological Resource Centres (BRCs)

Traditional culture collections are now being replaced by Biological Resource Centers (BRCs). This BRC-initiative was launched by the Organization for Economic Co-operation and Development. The OECD working Party on Biotechnology had its first conference in Tokyo in 1999 and the 'First Task Force' worked on the definition of Biological Resource Centers (BRCs). The report 'Underpinning the Future of life sciences and Biotechnology' (2001) deals with the establishment of national BRCs, the development of an accreditation system for BRCs based on international criteria, the creation of international linkages among BRCs, the coordination of standard rules and regulations and the establishment of a global BRC network (http://www.oecd.org/dataoecd/26/19/31685725.pdf) "Biological Resource Centers (BRCs) are an essential part of the infrastructure underpinning life sciences and biotechnology. They consist of service providers and repositories of the living cells, genomes of organisms and information relating to heredity and functions of biological systems. BRCs contain collections of culturable organisms (e.g. microorganisms, plant, animal and human cells), replicable parts of these (e.g. genomes, plasmids, viruses, cDNAs), viable but not yet culturable organisms, cells and tissues, as well as databases containing molecular, physiological and structural information relevant to these collections and related bioinformatics."

Crucial roles of BRC's include:

- Preservation and provision of biological resources
- R&D
- Conservation of biodiversity
- Repositories for protection of IPR
- Resources for public information and policy formulation

The evolution from Culture Collections to Biological Resource Centers will transform them from archives to treasuries of useful bioactive organisms serving the needs of industrial microbiology and biotechnology.

Assuming that natural products biodiscovery will remain an important part of new drug and bioactive molecules development in the new future, the capacity of culture collections should include the infrastructure, know-how and legal framework for biodiscovery. Their content in addition constitutes a huge wealth of organisms, most of which are still waiting to be studied for their bioactivity [33].

1.7. The Global Biodiversity Information Facility (GBIF)

The Global Biodiversity Information Facility is also an international initiative of high impact (http://www.gbif.org/). GBIF's mission is to make the world's primary data on biodiversity freely and universally available via the Internet (http://www.gbif.org/GBIF_org/what_is_gbif). Practically, GBIF encourages, coordinates and supports the development of worldwide capacity to access the vast amount of biodiversity data.

In the near term, GBIF will provide a global metadata registry of the available biodiversity data with open interfaces. Anyone can then use it to construct thematic portals and specialized search facilities. Building on the contents of this registry, GBIF will provide its own central portal that enables simultaneous queries against biodiversity databases held worldwide. In the long term, molecular, genetic, ecological and ecosystem level databases can be linked to the system. This will facilitate and enable data mining of unprecedented utility and scientific merit. As its work programs progress, GBIF will enable users to navigate and put to use the world's vast quantities of biodiversity information. This information is vital to generating economic, environmental, social and scientific benefits from the sustainable use, conservation and study of biodiversity resources.

In the long term, molecular, genetic, ecological and ecosystem level databases can be linked to the system. These will facilitate and enable data mining of unprecedented utility and scientific merit.

Resulting improved knowledge of the vast microbial diversity on earth will contribute positively to the worldwide bioeconomy through the discovery of many useful products it will continue to generate.

5. Conclusions

Despite its serendipitous nature, natural product screening has had a profound effect on the pharmaceutical, agrobiological and other biotechnological operations. The era of bioactive compound screening of secondary metabolites from microorganisms is being replaced by sophisticated, target-directed, mode of action screens [6, 41]. This approach relies on the technological tools of modern molecular biology and involves the search for low molecular-weight molecules with specific activities [7]. Concomitant with this new approach, knowledge of ecology and appreciation of biodiversity are used to direct microbial isolation programs, which in turn can influence the screen technology. In addition, understanding the role of secondary metabolism in nature provides valuable leads [7, 22].

It is becoming increasingly realized that evolution links groups of organisms in a way that has hitherto been scarcely recognized. Natural product screening has traditionally been regarded as a "needle-in-a-haystack" exercise, but recent research suggests that the search among microbes for molecules with receptor activity in mammalian systems has a previously unrecognized value and rationale behind it [25]. The success of such programs relies on the biodiscovery of microorganisms and the conservation of genetic resource they provide [6, 7]. Global coordination of culture collections is needed [35, 36]. Habitats need to be preserved and an international code of conduct needs to be established to bring about full cooperation and support between developing and developed nations so that the disciplines of ecology, taxonomy, and conservation can work alongside industry and the fields of molecular biology and biomedical science [6, 7, 25, 41].

References

1. Allen, T.W. and Ritchie, D.A. (1994). Cloning and analysis of DNA sequences from *Streptomyces hygroscopicus* encoding geldanamycin biosynthesis. *Molecular and General Genetics*, **243**: 593-599.
2. Allsop, A. (1999). Where the new classes of antibiotics and antifungals going to come from? *Microbiology Today*, **26**: 166-167.
3. Atlas, R.M. and Bartha, R. (Eds.) (1998). *Microbial Ecology, Fundamentals and Applications*. 4th Edition. Benjamin/Cummings Science Publishing, California, USA.
4. Browning, S. (2004). Money matters. *Microbiology Today*, **31**: 4-5.
5. Bull, A. T (1991). Biotechnology and Biodiversity. In *The Biodiversity of Microorganisms and Invertabrates: Its role in Sustainable Agriculture* (Hawksworth, D.L., Ed.). CAB International, Oxon, UK.

6. Bull, A.T. (Ed.) (2003). *Microbial Diversity and Bioprospecting*. ASM Press.

7. Bull, A. T., Ward, A.C. and Goodfellow, M. (2000). Search and discovery strategies for biotechnology: the paradigm shift. *Microbiology and Molecular Biology Reviews*, **64:** 573-606.

8. Egan, S., Wiener, P., Kallifidas, D. and Welligton, E.M.H. (2001). Phylogeny of *Streptomyces* species and evidence for horizontal gene transfer of entire and partial antibiotic gene clusters. *Antonie van Leeuwenhoek*, **79**: 127-133.

9. Fang, A.Q., Wong, G.K. and Demain, A.L. (2000). Enhancement of the antifungal activity of rapamycin by the coproduced elaiophylin and nigericin. *Journal of Antibiotics*, **53**: 158-162.

10. Fauth, U. (1996). New drug discovery from natural resources. In *Culture Collections to improve the quality of life* (Samson, R.A., Stalpers, J.A., van der Mei, D. and Stouthamer, A.H., Eds.). Ponsen & Looyen, Wageningen, The Netherlands, pp.405-408.

11. Fiedler, H-P. and Goodfellow, M. (2004). Alkaliphilic actinomycetes as a source of novel secondary metabolites. *Microbiology Today*, **25**: 27-29.

12. Frisvard, J.C., Bridge, P.D. and Arora, D.K. (Eds.) (1998). *Chemical Fungal Taxonomy*. Marcel Dekker.

13. Glowka, L. (1996). The Convention on Biological Diversity: Issues of interest to the microbial scientists and microbial culture collections. In *Culture Collections to improve the quality of life* (Samson, R.A., Stalpers, J.A., van der Mei, D. and Stouthamer, A.H., Eds.). Ponsen & Looyen, Wageningen, The Netherlands, pp.36-60.

14. Gürtler, H., Rassing, B.R., Hartjen, U., Braun, D.J. and Fels, J.J. (1996). Industrial exploitation of microbial diversity. In *Culture Collections to improve the quality of life* (Samson, R.A., Stalpers, J.A., van der Mei, D. and Stouthamer, A.H., Eds.). Ponsen & Looyen, Wageningen, The Netherlands, pp.402-404.

15. Hawksworth, D.L. (Ed.) (1991). *The Biodiversity of Microorganims and Invertebrates: Its role in Sustainable Agriculture*. CAB International, Oxon, UK.

16. Hawksworth, D.L. (1996). Microbial collections as a tool in biodiversity and biosystematic research. In *Culture Collections to improve the quality of life* (Samson, R.A., Stalpers, J.A., van der Mei, D. and Stouthamer, A.H., Eds.). Ponsen & Looyen, Wageningen, The Netherlands, pp. 26-35.

17. Hopwood, D. (2004). New drugs by manipulating *Streptomyces* genes. *Microbiology Today*, **31**: 64-65.

18. International Union for the Conservation of Nature and Natural Resources (1980). *World Conservation Strategy*. International Union for Conservation of Nature and Natural Resources, Gland.

19. International Union for the Conservation of Nature and Natural Resources (1988). *Convention on the Conservation of Biological Diversity*. International Union for Conservation of Nature and Natural Resources, Gland.

20. International Union for the Conservation of Nature and Natural Resources (1989). *From Strategy to Action*. The IUCN response to the Report of the World Commission on Environment and Development. International Union for Conservation of Nature and Natural Resources, Gland.

21. Kirk, O., Borchert, T.V., and Fuglsang, C.C. (2002). Industrial enzyme applications. *Current Opinion in Biotechnolology*, **13(4)**: 345-51.

22. Kurtböke, D.I. (Ed.) (2003). *Selective Isolation of Rare Actinomycetes*, Queensland Complete Printing Services, Nambour, Queensland, Australia.

23. Lazzarini, A., Cavaletti, L., Toppo, G., and Marinelli, F. (2000). Rare genera of actinomycetes as potential producers of new antibiotics. *Antonie van Leeuwenhoek*, **78:** 388-405.

24. Marrs, B., Delagrave, S. and Murphy, D. (1999). Novel approaches for discovering industrial enzymes. *Current Opinion in Microbiology*, **2**: 241-245.

25. Nisbet, L.J. and Fox, F.M. (1991). The importance of microbial diversity to biotechnology. In *The Biodiversity of Microorganisms and Invertabrates: Its role in Sustainable Agriculture* (Hawksworth, D.L., Ed.). CAB International, Oxon, U.K., pp.229-244.

26. Ogawa, J. and Shimizu, S. (1999). Microbial enzymes: new industrial applications from traditional screening methods. *Trends in Biotechnology*, **17**: 13-19.

27. Olembo, R. (1991). Importance of Microorganisms and invertebrates as components of Biodiversity. In *The Biodiversity of Microorganisms and Invertabrates: Its role in Sustainable Agriculture* (Hawksworth, D.L., Ed.). CAB International, Oxon, U.K., pp.7-15.

28. Omura, S. (1986). Philosophy of new drug discovery. *Microbiological Reviews*, **50**: 259-279.

29. Omura, S. (1992). Thom Award Lecture. Trends in the search for bioactive microbial metabolites. *Journal of Industrial Microbiology*, **10(3-4)**: 135-56

30. Palleroni, N.J. (1996). Microbial Diversity and the importance of culturing. In *Culture Collections to improve the quality of life* (Samson, R.A., Stalpers, J.A., van der Mei, D. and Stouthamer, A.H., Eds.). Ponsen & Looyen, Wageningen, The Netherlands, pp.111-114.

31. Primrose, S. and Gillard, R. (2004). Patent strategies for biotechnology companies. *Microbiology Today*, **31**: 6-7.

32. Riordan, C. (2004). Materials transfer agreements- "material" issues. *Microbiology Today*, **31**: 8-9.

33. Roblin, R.O. (1996). From collection-oriented research to collection-derived new products. In *Culture Collections to improve the quality of life* (Samson, R.A., Stalpers, J.A., van der Mei, D. and Stouthamer, A.H., Eds.). Ponsen & Looyen, Wageningen, The Netherlands, pp.202-205.

34. Sawada, K., and Ueda, M. (2001). Enzyme processing of textiles in reverse micellar solution. *Journal of Biotechnology*, **23**: 263-9.

35. Sly, L.I. (1996). WFCC: Aims and achievements revisited. In *Culture Collections to improve the quality of life* (Samson, R.A., Stalpers, J.A., van der Mei, D. and Stouthamer, A.H., Eds.). Ponsen & Looyen, Wageningen, The Netherlands, pp.3-16.

36. Sly, L.I. (1998). Australian Microbial Diversity. *Microbiology Australia*, **19**: 26-35.

37. Sosio, M. and Donadio, S. (2004). Exploiting and expanding actinomycete diversity. *Microbiology Australia*, **25**: 32-33.

38. Tan, R.X. and Zou, W.X. (2001). Endophytes: A rich source of functional metabolites. *Natural Products Reports*, **18**: 448-459.

39. Ten Kate, K. and Laird, S.A. (Eds.) (1999). *The Commercial Use of Biodiversity, Access to genetic resources and benefit-sharing*. Earthscan Publications Ltd., London.

40. Waites, M.J., Morgan, N.L., Rockey, J.S. and Higton, G. (Eds.) (2001). *Industrial Microbiology, An introduction*. Blackwell Science, London, UK.

41. Ward, A and Goodfellow, M. (2004). Taxonomy as a roadmap for search and biodiscovery. *Microbiology Australia*, **25**: 13-15.

42. Watve, M.G., Tickoo, R., Jog, M.M. and Behole, B.D. (2001). How many antibiotics are produced by the genus *Streptomyces*? *Archives of Microbiology*, **176:** 386-390.

43. Williams, S.T., Locci, R., Beswick, A., Kurtböke, D.I., Kuznetsov, V.D., Le Monnier, F.J., Long, P.F., Maycroft, K.A., Palmit, R.A., Petrolini, B., Quaroni, S., Todd, J.I. and West, M. (1993). Detection and identification of novel actinomycetes. *Research in Microbiology*, **144**: 653-656.

44. Woese, C.R. (2002). In *Biodiversity of Microbial Life: Foundation of Earth's Biosphere* (Staley, J.T. and Reysenbach, A.L., Eds.). John Wiley and Sons, Inc., New York, pp.xvi-xxii.

Intellectual Property Rights (IPR) for a Biological Resource Center as the interface between academia and industry

Y.-F. Chen and C.-C. Liao

Bioresource Collection and Research Center, Food Industry Research and Development Institute, 331 Shih-Pin Road, Hsinchu, Taiwan, ROC

1. Introduction

In an industrial structure like that in Taiwan, most companies belong to small and medium enterprise categories (SMEs) [17]. Because of the size restraints, SMEs generally lack research and development (R&D) capabilities. To upgrade the industrial structure of Taiwan by promoting industrial technology R&D, the government has set aside a budget to commission research institutes to participate in industrial technology R&D projects [10] Non-profit research institutes are important in the government's R&D efforts, and constitute an important bridge between the government and the private sector for technology transfer [16]. The Bioresource Collection and Research Center (BCRC) [9] was established as a multifunctional biological resource center (BRC) under these favorable circumstances, as a non-profit research institute supported by government funding for industrial technology R&D. Beside its culture collection and standard services, it endeavors to conduct industrial and technological research and is thus differentiated from other BRCs, especially from typical culture collections. Because of its active interaction with industry, BCRC can be presented as an example of a BRC playing a role between academia and industry.

To facilitate the commercial use of microbial resources, the transfer of tangible microbial resources and intangible technology is equally important [15]. In this respect, BCRC has first established a mechanism for transferring tangible microbial resources from providers to users in the academia and industry. Secondly, BCRC acquires intangible microbial technologies from the academic sector, protects technologies with intellectual property rights, and transfers technologies to the industrial sector.

2. Microbial resources: user/provider interface

2.1. Access and benefit sharing

The advance of biotechnology has accelerated significantly in recent years. Microbial resources have provided an important source in the development of biotechnology. Users of microbial resources, sometimes also the inventors of such technologies, can claim intellectual property rights (IPRs) to protect their invention [23] and can benefit from their IPRs. However, microbial resources have been treated as common human heritage that should be freely accessible [18]. Providers of microbial resources have thus been left behind in terms of financial benefit. Consequently, balancing the benefits of microbial resource providers and users is one of the main goals of the Convention of Biological Diversity (CBD) [4].

Catalyzed by the CBD, there has recently been a more regulated transfer of microbial resources than before. As required by Article 15 of the CBD, access to microbial resources (in short, access) should be regulated and fair and equitable sharing of the benefits arising from their utilization (in short, benefit sharing) should be considered. More specifically, access to microbial resources, particularly *in situ* [3] ones, will be subject to "prior informed consent" of competent providers, and benefit sharing will be based on "mutually agreed terms" between providers and users.

Along the microbial resources transfer chain, from an *in situ* resource origin to an *ex situ* [2] resource isolator, on to a BRC and to an *ex situ* resource user, the collection and provision for *ex-situ* microbial resources enable a BRC to monitor and track [20] the resource flow. Thus, a BRC plays a role in realizing the access and benefit sharing concept. Firstly, a BRC can encourage prior informed consent by requiring proof of the consent when acquiring microbial resources from a provider. Secondly, by using a material transfer agreement in a BRC, the bridge of the BRC can facilitate the creation of mutually agreed terms between providers and users of microbial resource.

2.2. General material transfer mechanism

BCRC has established a material transfer mechanism by adopting a standard material acquisition agreement (MAA) for the acquisition of microbial resources, and a standard material transfer agreement (MTA) for the provision of microbial resources (Fig. 1). MAAs enable microbial resource providers to define the terms they require, for example availability provision, intellectual property right provision, commercial

Figure 1: Transfer of Microbial Resources (MR) from providers to users through BCRC material transfer mechanism

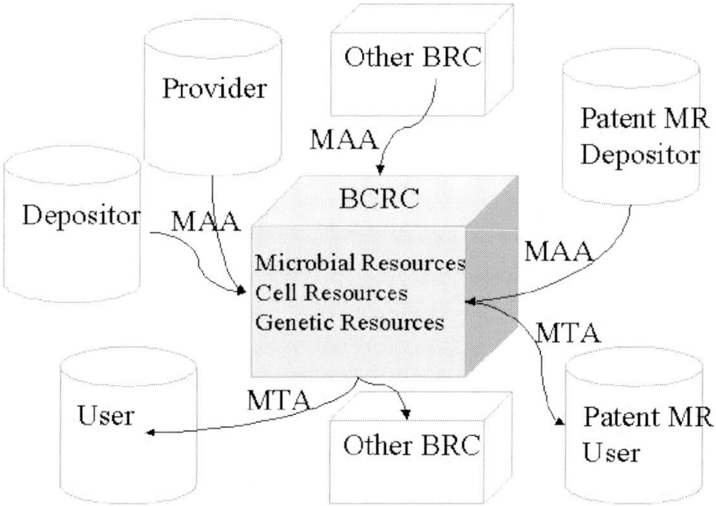

use provision, or benefit sharing provision. For availability provision, the provider can choose to make the microbial resources deposited in BCRC either: (1) publicly available without any restrictions (public deposit), or (2) not publicly available without their consent (safe deposit), or (3) only publicly available under certain terms and conditions (customized deposit), or (4) publicly available in accordance with the Taiwanese patent law (patent deposit, see section 2.3). Within MAA, the provider is encouraged to obtain prior informed consent from competent providers and is required to declare their authority to provide the microbial resources with the specified terms and conditions. Similarly, the MTAs of other BRCs when acquiring microbial resource from other BRC are treated as MAAs from the perspective of BCRCs.

Terms of each MTA faithfully correspond to the terms of each corresponding MAA. Thus, the mutually agreed terms between providers and users of microbial resources can be made possible thanks to the BCRC Bridge. Within MTA, further transfer of microbial resources to any third party is also prohibited. Limiting further transfer to third parties can shorten the transfer chain along which the monitoring of the transfer of microbial resources may be lost [5].

BCRC adopts MAAs and MTAs not only for access and benefit sharing purposes, but also to encourage the transfer of microbial resources between providers and users. From the point of view of the provider, the mechanism may secure the rights of the provider and thus encourage them to provide the microbial resources for distribution. Indeed, the number of deposits has increased significantly following the adoption of MAAs and MTAs in BCRC. The transfer of microbial resources from providers to users, for example between the academic and industrial sectors, is believed to be encouraged via the BCRC interface.

2.3. Patent material transfer mechanism

Apart from the agreement bound microbial resources, patent laws and regulations additionally regulate the patent-protected ones. For patenting purposes, some microbial resources are requested to be deposited in BRC to fulfill patent disclosure requirements [21]. The deposited microorganisms must then be made available to the public at certain times in accordance with the patent law. The rationale is consistent with the patent policy that encourages the disclosure of technologies to the public. Indeed, patent deposit is a mechanism that effectively forces the transfer of patented microbial resources, at least for R&D purposes [12].

For Taiwanese patent laws and regulations, biological materials that are not readily available to the public must be deposited at BCRC [11]. Several points should be noted regarding a deposition in Taiwan. First, depositions made under the Budapest Treaty [19] cannot fulfill the deposit requirement of Taiwanese patent law, because Taiwan is not a signatory to that Treaty. Thus, a domestic deposition in BCRC is necessary. Secondly, the depositions must be made before the patent filing date, as in most other countries [24]. Thirdly, the patent authority in Taiwan has promulgated a regulation for patent biological material deposit. BCRC performs the deposit in accordance with the regulations, unlike the contract relationship applied by some other depositaries [7]. Fourthly, the deposited materials are made publicly available upon granting the corresponding patent, unlike in some other countries [8]. The accessed deposited material can be used solely for experimental purposes, and cannot be transferred to any third party. Although some deposited materials may not be covered by the patent claims and thus their use is not bound by patent law, the accessed deposited materials are still limited by the regulations for experimental use only.

BCRC has established a security system specifically for patent deposit and transfer. Biological materials deposited in BCRC are treated confidentially and in a highly secure manner. Depositors can rely on the security system and the strict experimental use provisions included in the regulation to ensure the rights and flows of the deposited materials. This mechanism in BCRC is expected to encourage patent deposit and thus the transfer of microbial resources from the academia to industry.

2.4. Challenges and efforts

Besides the above mentioned transfer mechanisms, more challenges and efforts are needed to facilitate microbial resource exchange. Generally speaking, access to microbial resources, like other biological materials, is restricted in various ways, in terms of: (1) health and safety protection, (2) private industry protection, (3) import and export regulations, (4) IPR protections, and (5) MTA. BRCs should be made accessible to all legitimate users worldwide and should also serve as international gateways facilitating access to biological resources in the framework of international laws and regulations", as proposed by the OECD report [14].

BCRC is making various efforts to facilitate access in accordance with laws and regulations. In relation to (1) health and safety protection, laboratory facility information is required for users of microbial resources when applying for transfer. Users are requested and trained to examine the facility themselves to ensure that it provides a qualified experimental environment, reducing the risks of causing hazard to human, animal, plant, or environment. With reference to the (2) protection of private industry, through established material transfer mechanism, private industry sector collections are encouraged to transfer from BCRC to the public without losing their rights. For (3) import and export, laws and regulations are keeping collected. Permit application services and related consultation are provided to all those with such a requirement. As far as the (4) IPR protection is concerned, BCRC, as the patent depository of Taiwan, can transfer microbial resources thereby deposited in accordance with patent laws and regulations without problems. For microbial resources containing patented genes or gene fragments, IPR infringements are possible [22] in the process of transfer. BCRC plans to study this situation in the future. Suggestions based on this future study may be presented to the patent authority for consideration in policy-making [6]. Hopefully, a patent policy friendly to microbial resources transfer can be made. For (5) MTA, access restrictions to BCRC, when acquiring certain microbial resources, makes further transfer of resources impossible with BCRC. BCRC strives to improve its own authority nationally and internationally, partly by establishing a cooperative relationship. Thus, BCRC is expected to collect more microbial resources with the ability to transfer to the community in the future.

3. Microbial technology: academia/industry interface

To facilitate the commercialization of microbial resources, microbial technologies needed for industry are developed or acquired from academia into BCRC. Sources of technologies include: (1) in-house R&D, initiated and performed by BCRC, for those classified as important for the industry, (2) joint R&D performed by BCRC with cooperation and partial financial support from academia or industry, (3) academic R&D acquired by BCRC from academia and used as a basis for further R&D. The technologies are then protected by intellectual property rights, in the form of patents or trade secrets, for promotion. Ways of promoting technologies for commercialization in BCRC include: (1) technology transfer that usually accompanies the transfer of microbial resources; (2) contract tests based on BCRC established techniques, for example, bifidobacteria isolation; (3) open laboratory, which includes space and equipment that can be rented from BCRC for R&D purposes by companies; and (4) investment possibilities or spin-off for technologies of interest to BCRC. Examples are presented are below:

3.1. Model 1: Academia/BCRC/industry technology transfer

As shown in Fig. 2, BCRC acquires technology from academia and attempts to promote this technology to industry. In this particular example, the technology acquired involves a new genus of fungi with high infectivity towards pinewood nematodes that causes wilting of pinewood in eastern Asia. The fungi can be used as a biocontrol agent against nematodes to prevent loss of pine trees.

The new fungi isolated from nematodes in pine trees and named *"Estia vermicola"* was discovered and characterized by Prof. Tzean, an expert in phytopathology. BCRC identified his invention as having market potential. Meanwhile, Prof. Tzean was more interested in basic research than product development, besides he was not familiar with patent protection. Therefore, BCRC acquired the technology from Prof. Tzean and filed patent applications. Patents have been granted from some countries.

Furthermore, BCRC continues to conduct downstream R&D, including field testing, scale up production, and is also developing methods and formulas for administering

Figure 2: Academia/BCRC/Industry technology transfer case-nematophagous fungi

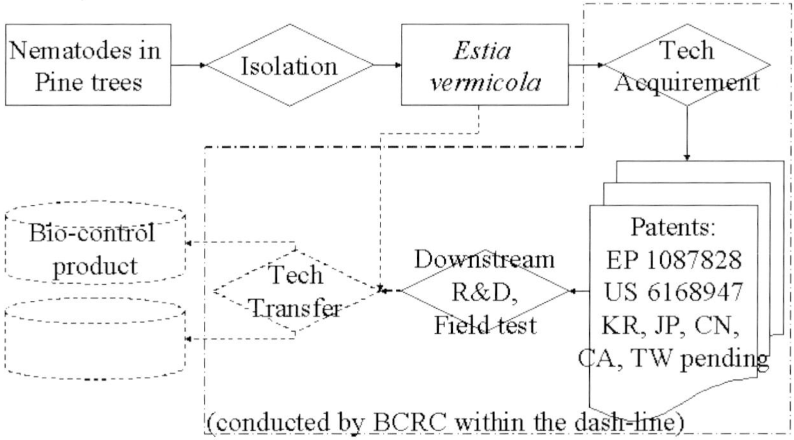

the fungi in pine woods. Further intellectual property protection concerning the related technology will also be followed. The technology package, including patents and know-how as well as the fungi, will be transferred to industry to produce a biocontrol product. Through a model such as this one, BCRC attempted to produce a more mature marketplace technology and to fill the gap between an academic discovery and a marketable product.

3.2. Model 2: BCRC/industry technology transfer

In the case displayed in Fig. 3, BCRC initiated a project, conducted its own R&D and transferred the technology to industry. The technology related to bifidobacteria, which are probiotics [1], can be used to make yogurt and many other functional foods. BCRC identified a need to keep the bacteria alive during storage in an aerobic environment, and also to keep the bacteria alive in human intestine under the stress of gastric acid and bile salt in stomach and intestine following consumption.

BCRC isolated new strains from infant faeces and improved the tolerance of these strains against bile salt, acid, and oxygen by mutation. The improved strains were then patented in some countries. Moreover, the patent and strains were then transferred to industry. The licensed company performed downstream processes to produce and market functional foods and snack foods using the new strains.

Figure 3: BCRC/industry technology transfer case-Bifidobacteria

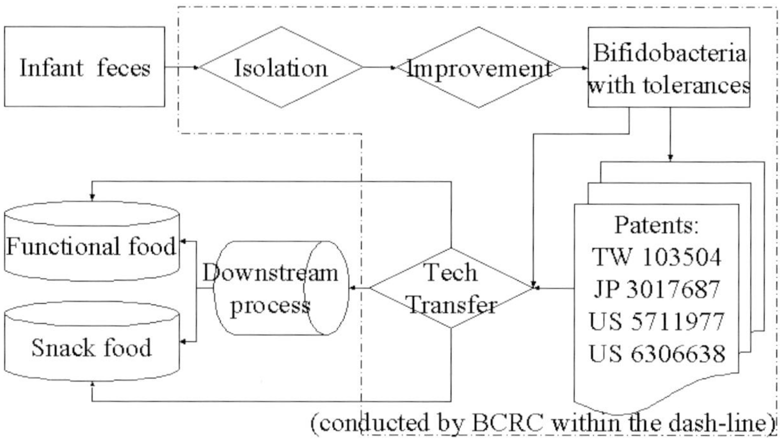

In another case, a technology was developed for improving traditional fermented foods. A mixed culture was isolated from traditional fermented tofu and formulated to allow the tofu to be produced on an industrial scale with higher quality and sanitation. The mixed culture was then patented and transferred to industry for commercialization.

Using a model like this in the present cases, BCRC has attempted to identify market needs and fill the gap between these needs and actual products.

3.3. Model 3: BCRC joint R&D with academia and industry

Fig. 4 shows another example involving joint R&D of BCRC with academia and industry. *Monascus* sp. strains are known to produce numerous valuable secondary metabolites, such as γ-aminobutyric acid (GABA) that can lower blood pressure, monacolin K that can lower cholesterol, and red or yellow edible pigment. BCRC identified the potential of *Monascus* sp. strains for developing drugs and food supplements.

BCRC isolated several strains from traditional fermented food. Some have conducted further improvement, and resulted in four valuable strains, one with high GABA productivity, one with high monacolin K productivity, one with red pigment, and one with yellow pigment. Patents were filed for the four strains. Besides, to make the strains more easily identifiable, each strain was given a consumer-friendly name for

Figure 4: BCRC joint R&D with academia and industry case-*Monascus* sp.

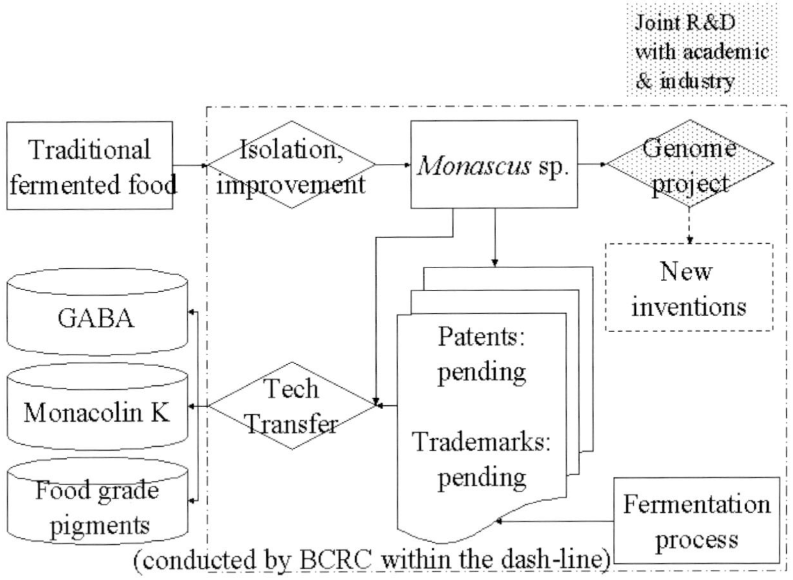

trademark registration. Notably, as in BCRC/industry technology transfer model, the strains and the subsequently developed fermentation process were transferred to companies for commercialization.

Furthermore, the strains are undergoing further R&D under a genome project with joint efforts from the academia and industry. The genome project includes whole genome sequencing, gene discovery, and gene manipulation. The goal is to discover new opportunities from the genome that embeds with abundant known and unknown valuable secondary metabolites. A strategic alliance is developing between the BCRC, academia and industry. Moreover, patent portfolios are also being constructed. Using a model as in Fig. 4 above, BCRC has been trying to integrate academic and industrial resources to identify additional pioneering areas to create new opportunities and establish further industries.

3.4. Model 4: BCRC open laboratory and investment

In the case illustrated in Fig. 5, BCRC rented laboratory space to and invested in a cord blood bank. A market has been emerging for umbilical cord blood collection recently because of the stem cells in the blood shedding light on regeneration therapy. A cord blood bank company has thus invited BCRC to participate in investment and technical support, and to provide laboratory space and equipment.

Firstly, BCRC provided its specialized technical support for preservation and bank management. Operating a cord blood bank, just like a BRC, requires preservation skills and bank management, including inventory and temperature control. Secondly, a laboratory and stock room space were rented to the company. Thirdly, monetary investment in the company was involved. Fourthly, the BCRC-developed medium for stem cell culturing was acquired by the company and a patent was filed.

The primary business of the company is to collect umbilical cord blood from new born infants in hospital, separate blood cells in the laboratory, and then store the blood cells in the bank. The company is also involved in R&D to provide new services, such as stem cell expansion, for clients with blood cells stored in the bank. The R&D work was not only conducted in cooperation with BCRC, but also used small business innovation research (SBIR) funds from government.

BCRC has been trying to support industry to complement their deficiencies by using models like those in the above cases.

Figure 5: BCRC open laboratory and investment case-Cord blood bank

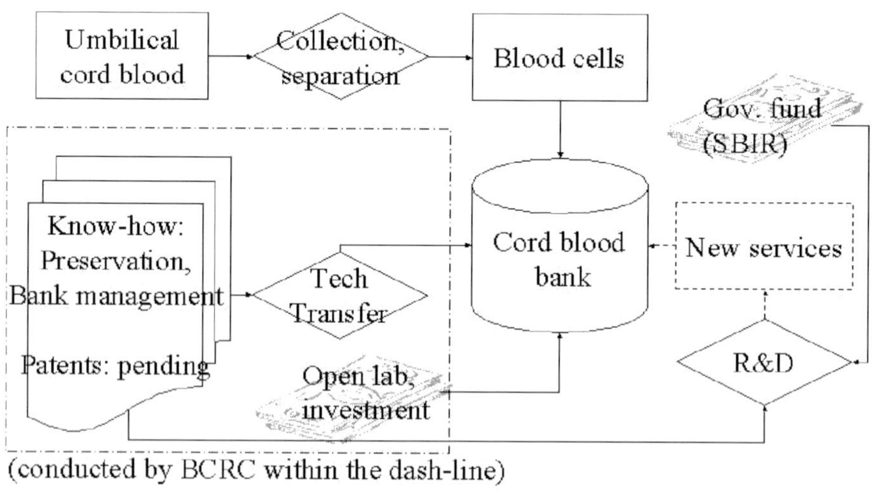

3.5. Challenges and efforts

Although facilitating the commercialization of microbial resources is considered an important mission in BCRC, the necessity for or outcome of this mission may be questioned. Being a non-profit research institute, BCRC is inevitably less market sensitive than the industrial sector. Determining the right direction for R&D to benefit companies needing help in industry is a challenge for BCRC. One of the main efforts undertaken by BCRC is to integrate R&D policy with IPR portfolios and industrial needs.

4. The role of intellectual property rights

4.1. Knowledge-based competitiveness

In this era of knowledge economy, IPRs have undoubtedly become enterprise assets. IPRs owners can make profits without building physical facilities, from letting the licensees establish physical facilities and make money for the IPR owners. IPRs build industry competitiveness.

On the other hand, IPRs represent the foundation of cooperation, and IPR protection encourages technology transfer. The benefits the IPR owners can gain from the licensees improve with licensee performance. IPR owners are thus willing to cooperate closely with licensees to achieve a win-win situation.

4.2. Incentives for microbial resources exchange

Without IPR protection, both individual scientists and industry would withhold these biological resources as "trade secrets" [14]. The protection of IPRs provides an incentive for microbial resource exchange under IPR mechanisms. Microbial invention requires a reliable patent microorganism depository. Like all International Depository Authorities under the Budapest Treaty. BCRC strives to establish a secure and reliable system for patent deposit purposes. Indeed, the patent depository of BCRC has been accredited by the ISO 9000 management system.

In a similar scenario, involving microbial resources but without any desire to seek IPR protection, the material transfer mechanism established by BCRC enables providers of microbial resources to secure their rights through a material transfer agreement. Moreover, IPR assignment provision can be placed in the agreement. The material transfer mechanism provides an incentive for microbial resources exchange under contract laws. Microbial resources exchange can thus be facilitated between academia and industry.

4.3. Incentives for microbial technology exchange

To complement R&D capability for some Taiwanese SMEs, BCRC through acquiring and promoting technologies fill the gaps between academic research and industrial commercialization. IPR protection is essential for all technology transfer in BCRC. In fact, both academia and research institutes like BCRC were not originally established for profit and were not designed or permitted to conduct technology commercialization. IPR protection is therefore the only way to obtain returns from the technology they developed. Using IPR protection, BCRC can not only secure its own technology, but also convinces academia to transfer technology downstream for commercialization.

5. Conclusion

IPR is closely linked to BRCs, particularly owing to it being the interface between academia and industry. First, patentees rely on the microbial resources depository of BRCs for patenting and non-patenting purposes. On the one hand, BRCs require IPR knowledge to operate the depository reliable for depositors. On the other hand, a reliable microbial resources transfer mechanism, such as MTA, also involves IPR provisions. BRCs as an interface between resources providers and users, and by using mechanisms for transferring both patent and non-patent microbial resources, can encourage and facilitate resource exchange. Secondly, BRCs may help fill the gap between technology in academia and industry. IPR protection is the only way for academia or a non-profit BRC allowing a return from self-developed technology. IPR provides incentives to and encourages academia or BRCs to transfer technology to industry.

IPR is only one of the steps required from BRCs. Numerous challenges still remain in facilitating the exchange of microbial resources and technologies. For example, restrictions affecting access to microbial resources must be managed to accelerate the exchange process. Monitoring and tracking of microbial resources is a priority for realizing access and benefit sharing, and consequently can encourage resource exchange. All challenges require efforts not only from individual BRCs, but also from the cooperation of BRCs, particularly on the international level.

Cited text

1. A "probiotic", by the generally accepted definition, is a "live microbial feed supplement which beneficially affects the host animal by improving its intestinal microbial balance". It has been argued that this effect of the probiotic on the intestinal ecosystem impacts in some beneficial way on the consumer.
2. According to Article 2 of CBD, "*ex-situ* conservation" refers to the conservation of components of biological diversity outside their natural habitats.
3. According to Article 2 of CBD, "*in-situ* conditions" means conditions where genetic resources exist within ecosystems and natural habitats, and, in the case of domesticated or cultivated species, in the surroundings where they have developed their distinctive properties.
4. At the 1992 Earth Summit in Rio de Janeiro, Convention on Biological Diversity was adopted. The Convention establishes three main goals: conservation of biological diversity, sustainable use of its components, and fair and equitable sharing of the benefits from the use of genetic resources.
5. BCCM, *op cit.*, para. I.5.
6. BCRC has been actively involved in research projects funded by the patent authority. These research projects included amending the guidelines for examining biotechnology patent in 2002. The guidelines have since been used by patent examiners in Taiwan.
7. For example, American Type Culture Collection.
8. For example, biological materials deposited for European patents can be accessed on the publication of patents without granting any patent rights.
9. For historical reason, BCRC has been a division of Food Industry Research and Development Institute (FIRDI), which is a non-profit research institute.
10. http://doit.moeaipo.gov.tw/newenglish/02_TDP/tdp_origins.asp
11. In the official announcement, "FIRDI" is the name of the depository. Moreover, BCRC, a division of FIRDI, is responsible for the depository.
12. Laws or regulations in certain countries such as Japan require that the accessed patent microbial materials can only be used for experimental purposes. However, some other countries, such as the US, do not pose any further use restrictions, provided no patent infringement occurs.
13. OECD, *Biological Resource Centers- underpinning the future of life sciences and biotechnology*, 2001, Chapter 5.
14. OECD, *op cit.*, p. 39.
15. One cannot improve upon a biotechnological invention without physical access to the germplasm, making exchanges of genetic material necessary. Quoted from Jong, S.C. and Cypess, R.H., (1998). Managing genetic material to protect intellectual property rights. *Journal of Industrial Microbiology and Biotechnology*, **20**: 95-100.
16. Small and Medium Enterprise Administration, Ministry of Economic Affairs, Taiwan, *op cit.*, p. 211.

17. SMEs accounted for 97.72% of all enterprises in 2003. (Small and Medium Enterprise Administration, Ministry of Economic Affairs, Taiwan, "White Paper on Small and Medium Enterprises in Taiwan, 2003", September 2003, p. 43.)
18. Sollerio, J.L. (1995). "Ownership of biodiversity: a developing country's perspective on an open international debate", In *Genes for the Future: Discovery, Ownership, Access,* (MacDonald, J.P. Ed.), National agricultural Biotechnology Council, Ithaca, NY, 1995, pp 109-116.
19. The Budapest Treaty on the International Recognition of the Deposit of Microorganisms for the Purposes of Patent Procedure was entered into force in 1980. The fundamental principle of the Treaty is that all parties recognize a deposit made in any one of International Depositary Authorities as sufficient for the purposes of their own patent procedure.
20. The concept of "monitoring" and "tracking" was proposed by BCCM, "micro-organisms sustainable use and access regulation international code of conduct", November 2000.
21. "The disclosure of the claimed invention shall be considered sufficiently clear and complete if it provides information which is sufficient to allow the invention to be made and used by a person skilled in the art as of the filing date, without undue experimentation..." (World Intellectual Property Organization, "Draft Substantive Patent Law Treaty", SCP9/2, 2003.) Where an invention depending on the use of microorganisms, a microorganism deposit might be used to fulfill the disclosure requirement.
22. The patent scope of gene patents does not necessarily cover microorganisms containing the claimed gene. Patent infringement must be analyzed on case by case basis, and is particularly relevant to the scope of the patented claims.
23. TRIPs 27.3b allows members to exclude "plants and animals other than micro-organisms" from patentability. A microorganism thus should be a patentable subject matter.
24. US patent law allows deposits to be made before issuing a patent.

Bioinformatics and Functional Genomics

Tsunami of Biological Data -Data resources and utilization-

H. Sugawara[1], T. Abe[1] and S. Miyazaki[2]

[1]Center for Information Biology and DNA Data Bank of Japan (CIB-DDBJ), National Institute of Genetics (NIG), Shizuoka 411-8540, Japan

[2]Laboratory for Information Biology, Department of Pharmaceutical Sciences, Faculty of Pharmaceutical Sciences, Tokyo University of Science, Chiba 278-8510, Japan

1. Explosion of biological data

Nucleotide sequences have been comprehensively deposited in the International Nucleotide Sequence Databases (INSD) that are composed of DDBJ (http://www.ddbj.nig.ac.jp/), EMBL database (http://www.ddbj.nig.ac.jp/) and GenBank (http://www.ncbi.nlm.nih.gov/Genbank/index.html). The size of INSD is over 30 giga-nucleotides in 30 mega-entries in December 2003. It has expanded 1.5-2 times a year even after the publication of the draft sequence of the human genome as shown in Table 1.

Table 1: Increase of the entries and base pairs in INSD

MM/YY	Entries	Nucleotides
03/04	32,693,678	38,008,449,840
12/03	30,405,173	36,079,046,032
09/03	27,753,140	34,280,225,489
06/03	25,149,821	32,162,041,177
03/03	23,250,813	29,711,299,332
12/02	20,354,812	26,931,456,316
09/02	18,401,358	22,782,404,136
06/02	17,260,693	20,158,357,982
04/02	16,503,157	18,579,627,226

Figure 1: The number of complete microbial genomes submitted to INSD by year

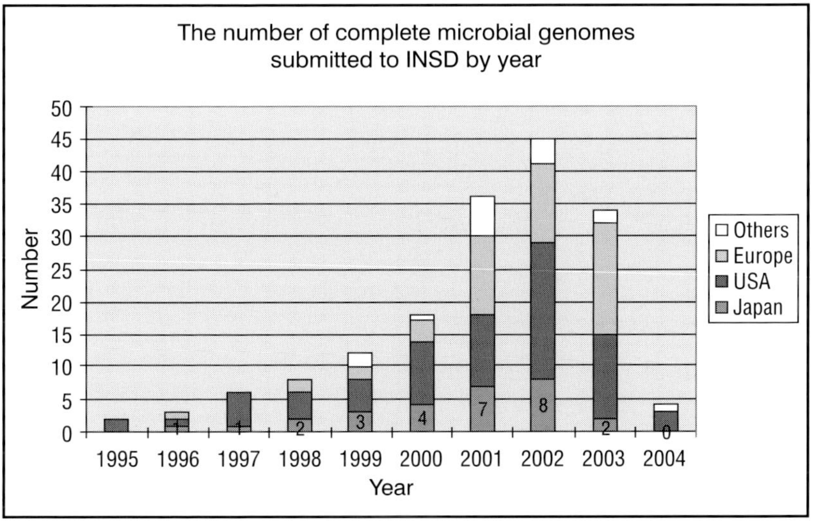

Among the INSD entries, the number of complete microbial genomes has very much increased in these years as displayed in Figure 1. It is to be noted that 70% of them were pathogens [16]. Even though the number decreased from the year 2002 to the year 2003, it is still significant and will remain significant. This is because agricultural, industrial and environmental microorganisms will be sequenced more and more in addition to pathogenic microorganisms.

A number of microbial genomes will actually be available in INSD according to the report at the Web site http://wit.integratedgenomics.com/GOLD/. The Web page compiles genome sequencing projects: completed but not yet disclosed, on-going and planned. The summary of the report is given in Table 2.

Table 2: Number of genome projects that will provide data to the public in the future

Region	US	Europe	Japan	Others
Archaea	11	4	2	1
Procaryote	168	63	12	14
Eucaryote	112	54	6	3

From Table 2, we can foresee that data of more than 100 procaryotic genomes will be available for the research and development in the near future. In addition, even the raw data from the whole genome shotgun that are not assembled into a complete chromosome will increase and be stored in the WGS division of INSD (http://www.ddbj.nig.ac.jp/).

The study of microbial diversity also contributes to the data explosion. Environmental sequences in the bacteria division of INSD have increased in these years to reach 79,916 entries by April 20th, 2004. The example for entry AB085364 is given in Figure 2.

The organism AB085364 in Figure 2 was not identified and is described as "uncultured bacterium" from "rumen contents". However, the product of a gene in the organism is inferred to be formyletetrahydrofolate synthetase. The entry AB085364 is an example of direct screening of useful genes from the environment. In 2004, Venter's group reported their accomplishment of applying whole genome shotgun sequencing to microbes in the Sargasso Sea. They identified 1.2 million new genes from at least 1800 genomic species [17].

In addition to nucleotide and translated amino acid sequences, a huge amount of data will be generated in proteome, transcriptome, metabolome and other "-omes". They will lead us to a new paradigm named omics. Considering the volume and variety of these biological data, it is obvious that the data science should be one of the key domains in biology in general. Even experts in biology need tools to mine data in order to: segregate signal from noise, squeeze information from data, and structure knowledge based on information. They will either intentionally or unconsciously apply tools of data mining to study living organisms and life phenomena.

2. Data science and polyphasic analysis

Are microbiologists novices in the data science? No, they are not. They have been heavy users of data analytical tools for many years. The numerical taxonomy based on phenotypic data [11] and the phylogenetic analysis [9] based on sequences of several genes have been common practices among microbiologists. The poly-phasic analysis based on multiple data and data analytical tools also becomes the common practice to understand the systematics of microbes [18].

In addition to phenotypes and sequence of genes, we are now able to perform comparative genomics to elucidate gene duplication, gene acquisition, gene loss,

Figure 2: The entry of environmental sequence in INSD

```
LOCUS       AB085364               1069 bp    DNA     linear   BCT 04-JUN-2002
DEFINITION  Uncultured bacterium gene for putative formyltetrahydrofolate
            synthetase, partial cds, clone:BPE01.
ACCESSION   AB085364
VERSION     AB085364.1
KEYWORDS    .
SOURCE      uncultured bacterium
            ORGANISM  Bacteria; environmental samples.
REFERENCE   1  (bases 1 to 1069)
     AUTHORS    Matsui,H., Tajima,K. and Ogata,K.
     TITLE      Direct Submission
     JOURNAL    Submitted (15-MAY-2002) to the DDBJ/EMBL/GenBank databases. Hiroki
                Matsui, Mie University, Faculty of Bioresources; 1515 Kamihama,
                Tsu, Mie Prefecture 541-8507, Japan
                (E-mail:
                        matsui@bio.mie-u.ac.jp, Tel:81-59-231-9640,
                Fax:81-59-231-9637)
REFERENCE       2
     AUTHORS    Matsui,H., Tajima,K. and Ogata,K.
     TITLE      Diversities of Formyltetrahydrofolate Synthetase Genes Recovered
                from Rumen Contents
     JOURNAL    Published Only in Database(2002)
COMMENT
FEATURES        Location/Qualifiers
     source     1..1069
                /clone="BPE01"
                /environmental_sample
                /isolation_source="rumen contents"
                /mol_type="genomic DNA"
                /organism="uncultured bacterium"
     CDS        <1..>1069
                /EC_number="6.3.4.3"
                /codon_start=2
                /product="putative formyltetrahydrofolate synthetase"
                /protein_id="BAB96947.1"
                /transl_table=11
                /translation="IGAANNLLAAMIDNHIYQGNELNIDPRKITWRRCVDMNDRQLRF
                VVDGLGGRVNGTPREDGYDITVASEIMAVLCLASDINDLKERPSRIIIGYTYGKVSEQ
                KPVTAGDLHAQGAMAALLKDALKPNLVQTLEHVPAIVHGGPFANIAHGCNSVTATRMA
                MKLADYAVTEAGFGADLGAEKFLDIKCRMAGIKPNAVVIVSTVRSLKYNGGVAKADLG
                PENLEALEKGLPNLLKHVSNIRDVYGLPCVFAINEFPTDTPAEVKLVEDRCKELGVNV
                VLSQVWAKGGEGGKALAEEIIRLVDQPSDFRFSYELDGTIEEKLNQIATKIYGARGVV
                MTADAAKQAKQLEAIGFGNCPI"
BASE COUNT  261 a     306 c        308 g        194 t
ORIGIN
      1  gatcggtgcg gccaacaacc tgctggcagc aatgatcgac aatcatatct atcagggaaa
     61  tgagctgaac atcgaccccc gcaagatcac atggagaaga tgcgtcgaca tgaacgaccg
    121  ccagctccgc tttgtggtgg acggactcgg cggaagagtc aacggaacgc cccgcgagga
    181  cggttacgat atcacggtag cgtctgagat catggctgtc ctgtgcctgg ccagcgacat
    241  taacgacctt aaagagagac cttccaggat catcatcggc tacacctacg gcaaagtcag
    301  cgagcagaag cccgtcacgg caggagacct gcacgctcag ggcgctatgg ctgccctcct
    361  taaggacgct ctcaaaccca atctcgtaca gaccctggag catgtccccg ctatcgtgca
    421  cggcggcccc ttcgcgaaca tcgctcatgg ctgcaactcc gtgacagcta caaggatggc
    481  tatgaagctg gcggattacg ctgttaccga ggcaggtttc ggcgcggatc tgggcgcgga
    541  gaagttcctg gacatcaagt gccgtatggc aggcattaag cccaacgcgg ttgtcatcgt
    601  gtccaccgtc agaagcctga agtacaacgg cggcgttgcc aaagcggatc tgggaccaga
    661  gaaccttgag gcgctggaga aaggtcttcc caaccttctc aagcatgtca gcaatatccg
    721  cgacgtgtac ggccttccct gcgtattcgc catcaatgag ttccccacag acactcctgc
    781  ggaagtgaag ctggtggagg acagatgcaa agagctcggc gtgaacgttg tcttgtccca
    841  ggtatgggca aagggcggcg aaggcggcaa ggccctcgcg gaggagatca tccgccttgt
    901  agaccagcct tcggacttcc gcttctccta tgaactggat ggaaccattg aggaaaaact
    961  caaccagatc gcaactaaga tctacggagc gcgcggagtc gtgatgacag ccgacgctgc
   1021  caaacaggca aaacagcttg aggcgatcgg cttcggaaac tgcccgatc
//
```

gene transfer, gene rearrangement, and the common ancestor. The minimum genome and complex phenomena in a cell may be inferable in a framework of the systems biology [7]. These are the new age poly-phasic analysis. In the 21st century, the poly-phasic analysis will evolve into a new stage thanks to omics.

In the following four sections, we will introduce search engines to find data sources, technologies that integrate heterogeneous data systems and the e-Workbench, which is a suite of data processing tools as well as a new data analytical tool.

2.1 Search engines to find data sources

The poly-phasic analysis is based on diverse data. The primary data repositories of microbial strains are the culture collections. However, strain data available on-line from culture collections are still limited. It is required to communicate with culture collections world-wide to retrieve data of authentic strains. Culture Collections world-wide compose the World Federation for Culture Collections (WFCC) that has WFCC-MIRCEN World Data Centre for Microorganisms (WDCM) as its component. WDCM (http://www.wdcm.org/) maintains the on-line database named CCINFO of the world directory of collections of cultures. The database CCINFO currently covers 476 culture collections in 63 countries. The user of CCINFO can search culture collections by acronyms, institute names, addresses, staff members, services available, categories of holdings, and scientific names of strains. The WDCM data management system is schematically displayed in Figure 3. The WDCM system is composed of 5 modules for: the registration and update by culture collections; maintenance of the master database by PostgresSQL; generation of reports; retrieval and browsing; and others. The WDCM system heavily uses XML and Java technologies [8], e.g., Servelet, XSP, XSL, and Xerces.

The STRAIN DB in Figure 3 is a list of scientific names and cell line names and is internally linked with the CCINFO database. Therefore, the user of the WDCM system can find culture collections by searching the STRAIN database. Distributed multiple collections are often hit by the search in CCINFO and STRAIN databases.

If you find several culture collections that seem useful, the WFCC HomePage at http://www.wfcc.info/ helps find out whether the culture collections maintain on-line databases at their Web sites. The WFCC HomePage provides the assembly of links to Web pages of culture collections. More detailed data can be accessed directly at the Web site by following links from the WFCC site. However, it will be tedious to visit all the sites one by one. Therefore, WDCM provides another search engine named AHMII (Agent to Help Microbial Information Integration) at http://www.wdcm.org/AHMII/ahmii.html [13] in three categories, namely, bacteria, fungi & yeasts and cell lines. AHMII in Figure 3 utilizes the threading function of the programming language of Java to carry out simultaneous parallel searches. The engine accepts a query from your Web browser and dispatches it to multiple Web sites distributed on the Internet.

Then the engine compiles the results of search at each site into an HTML file by first-come first-displayed base. It is to be noted that the Web sites are set by WDCM. In the categories of bacteria and fungi & yeasts, the taxonomy file is prepared and consistently used by INSD (http://www.ddbj.nig.ac.jp/intro-e.html). It means you can find some nucleotide sequence data of a microbe in INSD, if the scientific name of the microbe is included in the taxonomy file. You can also check the complete taxonomic lineage of the microorganisms there.

Figure 3: The WDCM data processing system

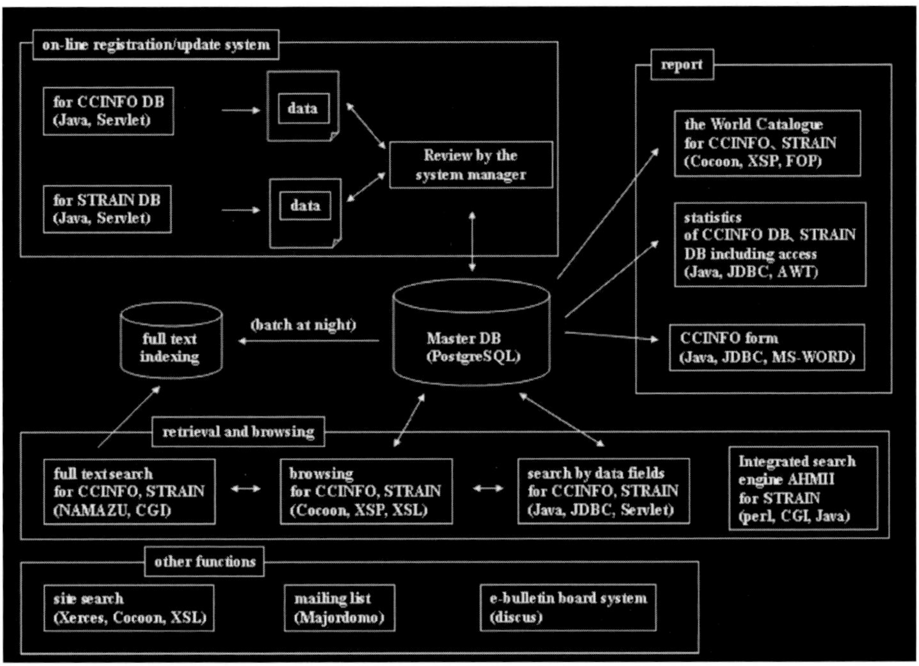

The WDCM site maintains not only the assembly of links to culture collection sites but also to other national, regional and global search engines, e.g., the UKNCC site of the national network of UK collections (http://www.ukncc.co.uk/), the CABRI site (http://www.cabri.org/) of Europe and the GBIF (Global Biodiversity Information Facility) site (http://www.gbif.org/). Namely, the WDCM site is a portal site for microbiology and culture collections.

2.2. SOAP and Web services

The sequence data of INSD and the databases of CCINFO and STRAIN are centralized databases that are robust and consistent. However, they cannot accommodate all types of databases and tools that evolve from time to time. It is important to have a system that can integrate data resources that are distributed on the Internet. The integration is essentially wrapping the heterogeneity of data systems. Since the late 20th century, new technologies for wrapping have been proposed, i.e., SOAP (Simple Object Access Protocol) and Web services. These technologies are standardized by the community of W3C (http://www.w3c.org/) and worth evaluating.

SOAP and Web services are for programmers and system engineers. The term "Web services" here does not mean services provided by conventional Web sites. It offers the interface of databases and data analytical tools to computer programs. We can let computer programs understand the specification described in Web services and mechanically utilize SOAP servers. The computer program will even be able to search Web services, if a universal directory of Web services is set up. Informaticians are able to program work flows that combine SOAP servers on demand in the near future. Then it will be feasible to develop a system that navigates the microbiologist to a work flow.

We actually implemented SOAP servers as the activity of the DNA Data Bank of Japan (http://www.xml.nig.ac.jp/) [14]. Since the size of INSD exceeded 30 million entries and 30 billion nucleotides in 2003, it is not feasible anymore to process a large number of entries and the huge amount of data by hand. The interface for direct access by a computer program to search, retrieve and analyze nucleotide sequence data is required.

SOAP servers at DDBJ provide the following services: getentry (get entries by specifying accession numbers, BLAST homology search, FASTA homology search, Smith–Waterman homology search, ClustalW multiple alignment, SRS (Sequence Retrieval System), TxSearch (retrieval of scientific names and lineage of an organism) and retrieval of DDBJ entries and extract features in annotation, the databases of GIB (Genome Information Broker) [4] and GTOP (Genomes TO Protein structures and functions) [6]. Computer programs written either in Perl or Java can call these services by just specifying the endpoint (URL address of a service) and the name of a method prepared in the service, e.g. if you want to retrieve all the sequence features in a certain region of a genome, the endpoint is http://xml.nig.ac.jp/xddbj/DDBJ/ and the method is getRelatedFeatures (number, star, stop) where the parameters in the parenthesis are the accession number of the sequence in DDBJ, and the start and stop position of the region to analyze.

The GIB (http://gib.genes.nig.ac.jp/) is a data repository of complete microbial genomes in the public domain which currently contains data on 18 strains of archaea, 141 strains of bacteria and 5 strains of eukaryota. In GIB, you can query any one strain or multiple strains by clone name, ORF name/number, function, gene name, product name, location and sequence. The result of query is displayed in graphics and also in a table format. The functions assigned to CDSs are: amino acid biosynthesis; biosynthesis of cofactors, prosthetic group and carriers; cell envelope; cellular processes; central intermediary metabolism; DNA metabolism; disrupted reading frame; energy metabolism; fatty acid and phospholipids metabolism; protein fate; protein synthesis; purines, pyrimidines, nucleosides and nucleotides; regulatory; signal transduction; transcription; transportation and binding proteins and others. RNA genes are classified into mRNA, rRNA, scRNA, snRNA, tRNA and others. Other sequence features such as RBS, promoter, intron, LTR and etc are also included, if available. If you want to get all the CDSs of *Escherichia coli* K12 MG1655 strain, the endpoint and method should be http://xml.nig.ac.jp/xddbj/Gib/ and getCDS ("Ecol_K12_MG1655:") respectively. Ecol_K12_MG1655 is the chromosome identifier uniquely assigned in GIB. It is possible to use GIB at http://gib.genes.nig.ac.jp/) interactively by hand with the Web browser and also interactively by computer programs with the SOAP interface. The same is true for the GTOP database.

2.2.1. e-Workbench for databasing, classification and identification

The polyphasic analysis is based on diverse data analytical tools. E-workbench, i.e. electronic workbench, is a general term and means the environment in which you can perform various types of data processing and carry out digital experiments. The e-Workbench for "databasing", classification and identification of microbes was named InforBIO [15], which is freely downloadable from the WDCM site (http://www.wdcm.org/). InforBIO offers a seamless operation from data management to data analysis. The user of InforBIO can design the database, import data from remote databases, export CSV files and utilize data analytical tools such as numerical taxonomy, phylogenetic analysis, discriminative identification and probabilistic identification. The functions of InforBIO are summarized in Table 3.

The user interface, the query system, data analytical programs and other utilities of InforBIO are written in Java, so that it is portable in WINDOWS, Macintosh and Linux machines. InforBIO uses XML [2] for the master file of the database to increase the portability of data among scientists. It is definitely convenient for a research community willing to use XML documents defined by the common DTD (Document Type Definition) or XML schema. The DTD of InforBIO incorporates elements for morphology, physiological data, biochemical data, genetic data, and sequence data.

Table 3: Functions of InforBIO

Fully open system by use of JAVA and XML
Integration of remote/local databases and tools
Import and export XML documents and CSV files
Numerical taxonomy
Primer design
Phylogenetic analysis
Probabilistic identification including construction of a standard data matrix
Identification by a decision tree including construction, import and export of a decision tree
Graphical representation of analytical results
Seamless operation from data retrieval to analysis and back
Available in CD-ROM

Look & feel of InforBIO is introduced in Figure 4. Phenotypic data and sequence data of the same set of microbes are analyzed by numerical and phylogenetic analysis respectively and the results are displayed in a dendrogram and a phylogenetic tree. It is easy to evaluate the consistency of the topologies of the trees in InforBIO by clicking the tabs of Dendro and ClustalW in Figure 4. Both the master file and sub-

Figure 4: Comparison of results of numerical and phylogenetic analysis in InforBIO (above: dendrogram, below: phylogenetic tree)

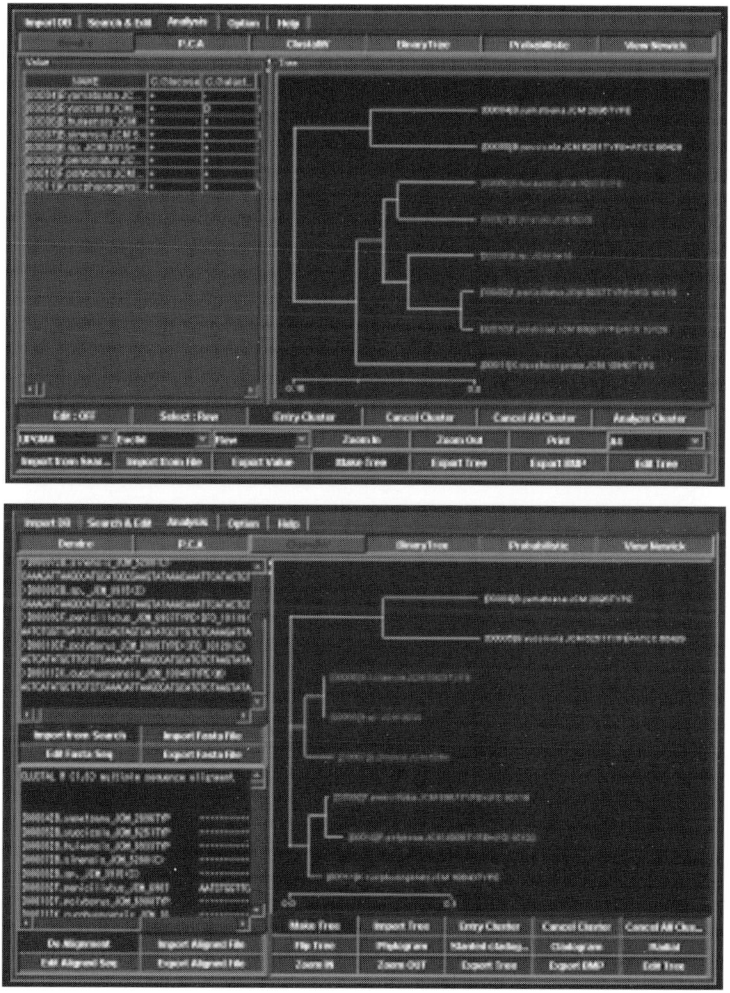

set of InforBIO are scalable to thousands of strains and high-performance servers in the Internet can be accessed from InforBIO for the analysis of a large data set.

The user interface, the query system, data analytical programs and other utilities of InforBIO are written in Java, so that it is portable in WINDOWS, Macintosh and Linux machines. InforBIO uses XML [2] for the master file of the database to increase the portability of data among scientists. It is definitely convenient for a research community willing to use XML documents defined by the common DTD (Document Type Definition) or XML schema. The DTD of InforBIO incorporates elements for morphology, physiological data, biochemical data, genetic data, and sequence data.

2.2.2. A new data mining tool – Self-organizing Map (SOM)

Metagenomic approach, which is genomic analysis of uncultured microorganisms, has been developed to identify novel and industrially useful genes and to study microbial diversity in a wide variety of environments [3]. In this approach, bioinformatic strategies to predict how many and what types of genomes are present in a sample are needed. In other words, an informatics tool to mine biological meaning from sequence data of fragments is a key to success. The data mining may also clarify how novel the genomic sequences are. To address such issues, one of the authors (TA) was involved in the development of application Kohonen's self-organizing map (SOM) to sequence data. [1, 5]. The modified SOMs are constructed for di-, tri-, and tetranucleotide frequencies in sequence fragments from a wide range of prokaryotic and eukaryotic genomes [1]. In the resulting SOMs, the sequences were clustered according to species without any information regarding the species, and increasing the oligonucleotides in length from di- to tetranucleotides increased the clustering power. SOM of tetranucleotide frequencies was improved for phylogenetic classification of genomic sequences from uncultured prokaryotes in environmental and clinical samples.

2.2.3. SOMs of prokaryotic genomes for which completesequences are available

Tetranucleotide frequencies of 81 prokaryotic genomes for which complete sequences were analyzed to evaluate the modified SOM. SOMs were constructed based on overlapping 5-kb sequences with a 1-kb sliding step resulting in Figure 5. These short sequences were tested because such fragments can be efficiently cloned and sequenced even for samples obtained from extreme environments.

In the DNA database, only one strand of a pair of complementary sequences is registered. The previous analysis of prokaryotic genomes showed that sequence fragments from a single genome are often split into two SOM territories that reflect the transcriptional polarities of the genes present in the fragment [1]; such a split was also observed in the Tetra-SOM in Figure 5. For phylogenetic classification of sequences from uncultured prokaryotes, it is not necessary to know the transcriptional polarity in the sequence fragment and the split into two territories complicates assignment to a species. Therefore, we tested a new type of SOM in which a pair of complementary tetranucleotides (e.g., AATC and GATT) was added, and the SOM for the degenerate sets of tetranucleotides was designated DegeTetra-SOM. This approximately halved the computation time, and the clustering power was higher than that of the Tetra-SOM. In the phylogenetic classification of uncultured prokaryotes, especially for those found in novel or extreme environments, the classification into major phylogenetic groups rather than into individual species is important. Therefore, the classification of the 226,000 sequences of known prokaryotes into 11 major phylogenetic groups was carried out (Group in Figure 5); nodes that contain sequences from one phylogenetic group are indicated in color, and those that contain sequences from more than one group are shown in black. The number of black nodes, which contained sequences of more than one group, decreased significantly in comparison to that of the species classification. Approximately 88% of 5-kb sequences were classified into the appropriate group territory in the DegeTetra-SOM. Phylogenetically related species often had neighboring territories and, therefore, produced united territories in the group classification. Interestingly, even in the united territories, species borders were often visible as continuous white nodes that contained no genomic sequences in the 5-kb SOMs. The vectors of species-specific nodes, even near a species-territory border, were distinct between species, and species borders could be drawn automatically on the basis of the contiguous white nodes.

2.2.4. Sequences from species-unidentified prokaryotes

One goal of metagenomic studies is to reconstruct individual genomes of uncultured microorganisms by sequencing a large amount of the genomic DNAs contained in metagenomic libraries derived from an environmental sample [10]. Such technology should assist in developing accurate views of the ecology of environmental microorganisms and allow extensive surveys of sequences useful in industrial and scientific applications. However, with a simple compilation of a large number of sequence fragments, it appears to be impossible to predict what kind of species and the ratios of species present in an environmental sample, to which lineages the species belong, and how novel the genomes are. SOMs can be the solution.

As for the first practical application of the proposed method, non-rDNA sequences from unidentified or uncultured prokaryotes in INSD were characterized; sequences

Figure 5: Tetra- and DegeTetra-SOMs for overlapping 5-kb sequences of 81 genomes of 65 prokaryotic species

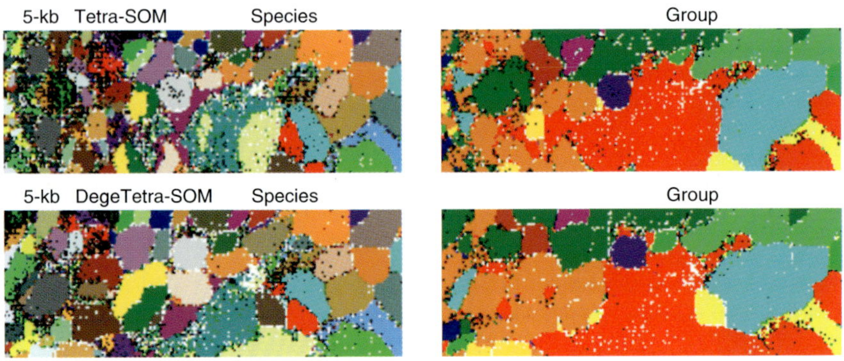

Species: sequence classifications into each species. Nodes that contain sequences from more than one species are indicated in black, and those containing sequences from a single species are indicated in color (data not shown). Sequences of closely related prokaryotes such as different strains of one species are indicated by the same color.

Group: sequence classifications into 11 phylogenetic groups. Nodes that contain sequences from more than one group are indicated in black, and those containing sequences from a single group are indicated in color as follows: α-Proteobacteria (■), β-proteobacteria (■), γ-proteobacteria (■), δ-proteobacteria (■), Archaea (■), Chlamydia (■), Firmicutes (■), Actinobacteria (■), Fusobacteria (■), Thermotogae (■), and others (■).

were submitted directly to the DDBJ/EMBL/GenBank often without publication in literatures. The number of 660 non-rDNA genomic sequences longer than 1 kb was found in INSD and allowed the classification of the 660 sequences into 11 prokaryotic groups in the following way. DegeTetra-SOM was constructed in advance with nonoverlapping 5-kb sequences from the 147 prokaryotic genomes for which almost complete sequences are currently available (Known prokaryotes in Figure 6A). For this SOM, instead of the 81 genomes used in Figure 5, 147 genomes were used to analyze a wider range of sequences because sequence redundancy due to inclusion of closely related genomes, such as those of different strains of one species, is not

Figure 6: SOMs for phylogenetic classification of sequences from species-unidentified prokaryotes

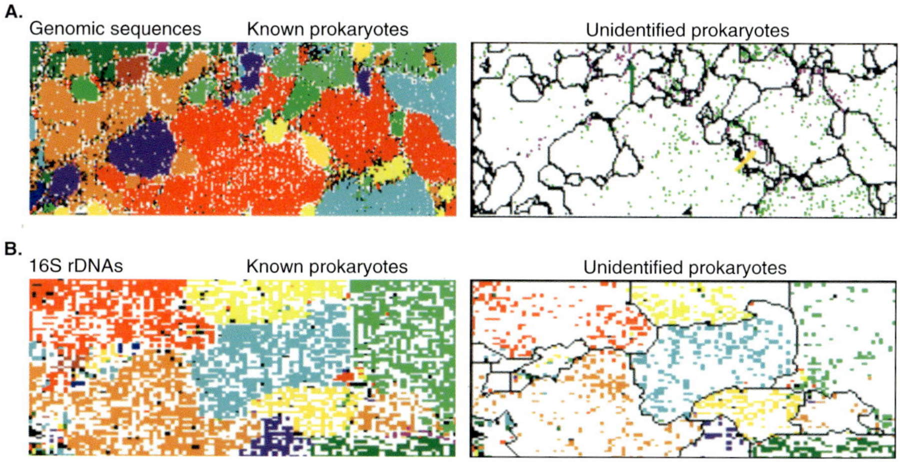

(A) 5-kb DegeTetra-SOM of 147 prokaryotic genomes. Known prokaryotes: sequence classification into 11 phylogenetic groups for species-known prokaryotes. Nodes are marked as described in the group classification in Figure 5. Unidentified prokaryotes; 660 non-rRNA sequences from species-unidentified prokaryotes were mapped on the 5-kb DegeTetra-SOM. The 343 sequences from rumen contents submitted directly to DDBJ/EMBL/GenBank by Matsui *et al.* (AB085236-AB085579) are specified in pink.

(B) Tetra-SOM of 16S rDNAs from known prokaryotes. Known prokaryotes: classification into 11 phylogenetic groups. Nodes are marked as described in the group classification in Figure 1; others were not included. Unidentified prokaryotes; rDNA sequences from species-unidentified prokaryotes, for which the group classification is annotated in DDBJ/EMBL/GenBank, were mapped on the 16S rDNA Tetra-SOM. Nodes where sequences of a single group were mapped are colored showing the group and those containing sequences of more than one group are marked in black.

problematic for the group classification. Nodes that contained sequences from one group are marked in color, and those that contained sequences of more than one group are shown in black. Although the pattern became slightly more complex than that of the 81 genomes, classification according to the group was again apparent. The 660 non-rDNA sequences from species-unidentified prokaryotes were then mapped on this 5-kb DegeTetra-SOM (Unidentified prokaryotes in Figure 6A). Detailed inspection of the sources of these non-rDNA sequences with reference to DDBJ annotations revealed that approximately 50% corresponded to sequences obtained from rumen contents. These sequences are specified in pink. There was a characteristic zone composed primarily of the respective sequences (11% of sequences from the rumen contents; specified with a green arrow in Unidentified prokaryotes in Figure 6A) in an Archaea territory that contained *Methanobacterium, Methanosarcina* and *Thermoplasma* sequences. This indicated that the rumen sequences were derived from Methanogens. Another characteristic clustering for the rumen sample was in the d-proteobacteria territory containing *Desulfovibrio desulfuricans* (8% of the rumen sequences; a dark yellow arrow), indicating the sequences were derived from sulphate-reducing bacteria. The anaerobic species are expected to be present in the rumen. When genome sequences of a wide range of taxonomic groups accumulate, SOM will become a wide-applicable and powerful tool for phylogenetic classification of genomic sequences, which is most useful for classification of sequences obtained from a mixture of uncultured microorganisms [17].

4. Concluding remarks

We first introduced that the tidal wave (tsunami) of sequence data has been caused by genomics and biodiversity studies. We also pointed out that there is a huge amount of data off-line, e.g., data maintained in culture collections. Then we introduced search engines either to approach the data source or capture actual data, e.g.: **WDCM** (WFCC-MIRCEN World Data Centre for Microorganisms) to search culture collections and their holdings; **GIB** (Genome Information Broker) for the retrieval and analysis of genome sequences and their annotation; and comparatively new information technologies for the integration of diverse databases and data analytical tools, namely, **XML**, **SOAP** and **Web services**. We believe that the Web services will be explored by many bioinformatics groups [12, 19]. It will be quick and easy for bioinformaticians to find and call the service from their programs, even though biological data resources keep evolving.

Once you have various categories of data, you certainly need a suit of application programs to analyze data. The e-workbench named **InforBIO** is an example of a suite. Work flows based on Web services provided by INSD, culture collections, publishers and other public servers will be another solution. Furthermore, we have to develop and apply new algorithms to the tsunami of biological data. The **SOM**

(Self-Organizing Map) is one of promising tools to retrieve biological meaning from raw data. It can not only be applied to not only sequence data but also to other types of data such as expression data. Ultimate utilization of the tsunami of biological data is to understand evolution, development and homeostasis of organisms as an integrated system. We need more data from *omic*s, more interoperability among servers and more advanced data mining tools to reach the ultimate goal.

Acknowledgements

We thank Mr. Yasumasa Shigemoto and Mr. Mashito Yamaguchi of Fujitsu for their support in the development of SOAP servers and WSDL. We are also grateful to the Ministry of Education, Science, Culture, Sports and Technology and also the Japanese Science and Technology Agency for their enduring financial support.

References

1. Abe, T., Kanaya, S., Kinouchi, M., Ichiba, Y., Kozuki, T. and Ikemura, T. (2003). Informatics for unveiling hidden genome signatures. *Genome Research,* **13**: 693-702.
2. Achard, F., Vaysseix, G. and Barillot E. (2001). XML, bioinformatics and data integration. *Bioinformatics,* **17**: 115-125.
3. Amann, R.I., Ludwig, W., and Schleifer, K.H. (1995). Phylogenetic identification and in situ detection of individual microbial cells without cultivation. *Microbiology Reviews,* **59**: 143-169.
4. Fumoto, M., Miyazaki, S. and Sugawara, H. (2002). Genome Information Broker (GIB): data retrieval and comparative analysis system for completed microbial genomes and more. *Nucleic Acids Research,* **30**: 66-68.
5. Kanaya, S., Kinouchi, M., Abe, T., Kudo, Y., Yamada, Y., Nishi, T., Mori, H. and Ikemura, T. (2001). Analysis of codon usage diversity of bacterial genes with a self-organizing map (SOM): characterization of horizontally transferred genes with emphasis on the *E. coli* O157 genome. *Gene*, **276**: 89-99.
6. Kawabata, T., Fukuchi, S., Homma, K., Ota, M., Araki, J., Ito, T., Ichiyoshi, N. and Nishikawa, K. (2002). GTOP: a database of protein structures predicted from genome sequence. *Nucleic Acids Research*, **30**: 294-8.
7. Kitano, H. (2002). Systems Biology: A Brief Overview. Science, **295**: 1662-1664.
8. Maruyama, H., Tamura, K., Uramoto, N., Murata, M., Clarl, A., Nakamura, Y., Neyama, R., Kosaka, K. and Hada, S. (2002). *XML and Java* (Second Edition): *Developing Web Applications*. Pearson Education Japan, Tokyo.
9. Nei, M. (Ed.) (1987). *Molecular Evolutionary Genetics*, Columbia University, New York.
10. Schloss, P.D., and Handelsman, J. (2003). Biotechnological prospects from metagenomics. *Current Opinion in Biotechnology,* **14**: 303-310.
11. Sneath, P.H. and Sokal, R.R. (1973). *Numerical Taoxnomy: The Principles and Practice of Numerical Classification*. W.H. Freeman, San Francisco.

12. Stein, L. (2002). Creating a bioinformatics nation. *Nature,* **417**: 119–120.

13. Sugawara, H. and Miyazaki, S. (2003). (1). AHMII: Agent to Help Nmicrobial Information Integration. *Nucleic Acids Research,* **31**: 3727-3728.

14. Sugawara, H. and Miyazaki, S. (2003). (2). Biological SOAP servers and web services provided by the public sequence data bank. *Nucleic Acids Research,* **31**, 3836-3839.

15. Sugawara, H., Tanaka, N. and Miyazaki, S. (2003). An e-Workbench for the Study of Mcirobial Diversity: the System Design and Basic Functions. *Microbiology and Culture Collections,* **19**: 59-67.

16. Tanaka, N., Hirahata, H., Miyazaki, S. and Sugawara, H. (2004). The status quo of microbial genomic data available in the public domain: archaea and bacteria 2003, *WFCC Newsletter,* **38**: 12-18.

17. Venter, J.C., Karin, R., Heidelberg, J.F., Halpern, A.L., Rusch, D., Eisen, J.A., Wu, D., Paulsen, I., Nelson, K.E., Nelson, W., Fouts, D.E., Levy, S., Knap, A.H., Lomas, M.W., Nealson, K., White, O., Peterson, J., Hoffman, J., Parsons, R., Baden-Tillson, H., Pfannkoch, C., Rogers, Yu-Hui, R., and Hamilton, O.S. (2004). Environmental Genome Shotgun Sequencing of the Sargasso Sea. Science, **304**: 66-74.

18. Vandamme, Peter A.R. (2001). Polyphasic Taxonomy in Practise: the *Burkholderia cepacia* Challenge. *WFCC Newsletter,* **34**: 17-24.

19. Wilkinson, M.D. and Links, M. (2002). BioMOBY: an open source biological web services proposal. *Briefings in Bioinformatics,* **3**: 331–341.

Bioinformatics: Perspectives for a wider network

B. Pot

Applied Maths B.VB.A., Keistraat 120, B-9830 Sint-Marten-Latem, Belgium

1. Introduction

The value of any center for biological resources is directly linked to the amount and quality of the information associated with each collected specimen. As an example a bacterial culture collection can be considered. Besides the most commonly accepted need to describe source and history of a culture, its microbiological potential is important to register. This could involve genetic, phenotypic, biochemical and taxonomical information. The possibility to study and detect links between these different sources of information can add an additional dimension to the value of a collection and increase its potential for valorization. At this stage the bioinformatic tools to physically link data from different origins or to combine them in a single integrative approach are not readily available.

On top of this, the need to efficiently capture, store, process and analyze (micro)biological information has increased over the last two decades. Not only did the technologies involved become more performing, yielding increasing amounts of information, (e.g. large scale sequence analyses, micro-array analyses, proteomic approaches), also the number of techniques yielding various types of information has exploded: many new typing techniques based on PCR and / or restriction analysis or based on sequences (ARDRA, AFLP, RAPD, ERIC-, BOX-, REP-PCR, PFGE; T-RFLP, DGGE, TGGE; MLST, MLVA, VNTR, Microsats, etc...), MALDI & SELDI techniques in the proteomics field, protein-arrays, labs on a chip, etc...). Over the years, large databases, public and private, harboring much of the basic information necessary to support these techniques, are made available for worldwide access over the Internet (NCBI, EMBL, SWISSPROT, KEGG, OMIM, Gene Ontology, etc...).

However, the aspect of **quality control** of this type of information has long been neglected, but may actually proof to be determinative for the reliability and acceptability of the outcome of any analysis using these available 'public' data.

Flexibility of software trying to link up with numerous public databases or allowing to search for 'public' information (e.g. a sequence BLAST) offers little benefit if the quality of the searched database is below the quality of your own information. Therefore complex quality monitoring tools are needed which use different data sources to validate or interpret the result of such a data search.

This again stresses the need to **link different types of information**, e.g. compare gene expression data measured by micro-array analysis to protein expression values measured by 2D protein electrophoresis. This can easily be organized through a linkage with relevant pathways involved and could confirm the quality of the approach or the data obtained. When on top of this, genome or chromosome information is available, regulation aspects or co-expression can further be validated by structural chromosome analysis revealing positions of the genes on the chromosome and confirming a hypothesized linkage.

Especially when considering genome wide information, **computer power** may become a critical aspect, as well as the database potential needed. Fortunately, in the last twenty years, elevated computing - and storage capacity became available at democratic prices, introducing a high potential for professional data storage and data analysis in virtually any routine and research laboratory as well as in any industrial setting. The widespread storage and analysis of this information has been very rewarding for research purposes (data exchange, ease of interpretation, etc), for industries (in terms of quality control, response time, etc…) and could surely improve the communication with the government. For all these purposes simple visualization of complex analytical results requires again the use of specific tools. Self-organizing maps can, as an example, very easily represent a complex dataset containing tens of thousands of entries, a representation that is e.g. no longer feasible by traditional dendrogram or PCA representation.

This paper discusses the potential benefits of uniform software packages to improve interaction between information sources derived from different suppliers, with a potential to process, store and analyze this diversity of data. Considerable attention is given to the potential problems and pitfalls.

2. The state of the art: hurdles to be taken!

Related to the process of democratization of analysis power, the problem of **data exchange and compatibility** has increased. Compatibility problems increased with expanding bio-informatics-diversity: many academic institutions used public and private funding to build their own bioinformatics environment. The solutions they developed were often solving a local problem or improving an existing approach, but were rarely structured to match minimal criteria for data exchange or for wider distribution. In these applications interfaces to other bioinformatics tools are often limited or non-existing. Therefore, one will find today on the Internet a wealth of bioinformatics tools that rarely are adapted to a particular data type nor prepare the output in a standard format. Therefore, a successful solution to the processing of biological data will most probably be offered by the use of a limited number of bioinformatics tools, which have the potential to cover a wide range of possible biological applications. Given the complexity and diversity of the data generated today, a final solution will need to be implemented as a modular structure that can easily be expanded to cover varying needs. In such a structure, statistical and mathematical processing tools will link seamlessly to database- and communication tools, offering full functionality packed behind a uniform interface.

Integration of data from different sources requires **high levels of standardization**. This standardization is required on the hardware level (e.g. correctness of different types of sequencers to determine a particular fragment length; see ENEMTI report on http://arch.rivm.nl/enemti/Scientific-%20Report%20corr%28121201%29.doc) but also on the software level. Software can easily help to overcome hardware differences: by including reference samples at regular times, the hardware performance can be calibrated with preset standards and can be corrected if necessary.

Software standardization also covers e.g. the consistent use of uniform accession numbers, enzyme numbers, sequence annotations, embedded in or retrieved from standard formats for e.g. sequences (EMBL, GenBank, DDBJ), proteins (SWISSPROT) or micro-arrays (MIAME format for data in ArrayExpress, GEO, CIBEX). For many other data types these standards are not yet available: fingerprint profiles (images or curve formats), phenotypic characters (API, BIOLOG, FAME, VITEK, and so on), or simply the descriptive information. For these data types, storage in a more widely accepted database (Microsoft® SQL Server™, Oracle®, Microsoft® Access, PostgreSQL, MySQL, and so on) with uniform import and export potential (Standard Query Language, ODBC and so on) may improve the ease of exchange.

The construction of an integrated database relies entirely on links to external information sources and on regular updating routines. Data exchange formats such as ODBC or XML gained a lot of interest over the last years since they allow overcoming structural differences in communication protocols for data types of a variety of origins.

The **quality of the data** has not escaped the 'Gaussian distribution'. Many laboratories manage to deliver high quality data, but for some the use of less performing hardware or software has resulted in degenerated data. As a consequence public databases will contain good quality data but will be 'contaminated' to some extend with less reliable or even questionable data. Modern bioinformatics tools therefore need to allow the evaluation of data quality as well. Many different strategies are possible. The most logic way is to seek confirmation from different, independent sources, or try to correlate different types of information in a more broad integrative approach, diluting the possible negative influence of one partially 'contaminated' dataset.

Other simple techniques can also be used. When multiple 16SrRNA sequences of a single bacterial species are available, only very rarely these sequences should differ for more than 4 to 5 percent, (the generally accepted difference on the species level is 3% [1]). For good quality sequences, two directions (3'-5' and 5'-3' strands) should be requested and incomplete or partial sequences can be refused. For micro-array data, p-values can be calculated based on duplicate probes on the same chip. In addition, the performance of well-chosen control probes applied on each chip can be checked over time. Statistical methods, integrated in the software, are indispensable to evaluate these controls. When multiple data sets are available, it is even possible to calculate the congruence between these data, trying to restrict further analyses to the most interesting approach(es) (Figure 1). Also the consistent use of well-chosen and uniform reference systems for matching new data before storage is considered will improve considerably the value of any database constructed. To that extend, proper tools need to be used to detect and correct for inevitable experimental variations (Figure 2). When exchanging information between different institutes the lack of identical reference systems may be a serious bottleneck. Figure 3 shows one way, through 'remapping', to compare fingerprints run with different reference systems.

Figure 1: Congruence between similarity matrices obtained for a diversity of different techniques, performed on the data set from Figure 4

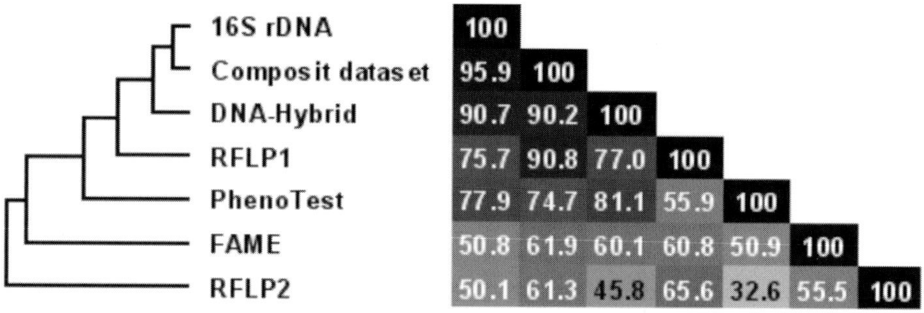

16SrDNA: comparison of multiple aligned DNA sequences, coding for the 16S ribosomal RNA (average length of 1300 nucleotides); DNA-Hybrid: result of DNA:DNA hybridizations (optical renaturation method) of fragmented chromosomal DNA's; only about 10% of the possible hybridizations have been performed; RFLP1: restriction fragment length polymorphism patterns with one restriction enzyme (see Figure 4), compared by the Pearson Product Moment Correlation Coefficient (curve – based comparison); RFLP2: restriction fragment length polymorphism patterns with a second restriction enzyme, compared with the Dice coefficient (band – based coefficient); PhenoTests: results of a API20 strip for all strains listed in Figure 4, compared by the Pearson Product Moment Correlation Coefficient; FAME: results of a gaschromatographic analysis of the collection studied (MIDI system), compared by the Euclidean Distance Coefficient ; Composit dataset: calculated combination of all previous dataset, representing a 'consensus' grouping, based on the average similarity for each couple of strains and correcting for the number of characters in each test.

(Software used: The BioNumerics® Suite, version 4.0; Applied Maths B.V.B.A., Sint-Martens-Latem, Belgium)

Figure 2: In Lane 1 of this AFLP fingerprint gel, the software has detected a possible displacement of the bands of a reference profile versus a reference system, shown on the left. A correction (see arrows pointing up) will be performed to align and straighten the reference trace, as well as all neighboring data lanes (lanes 2 to 6).

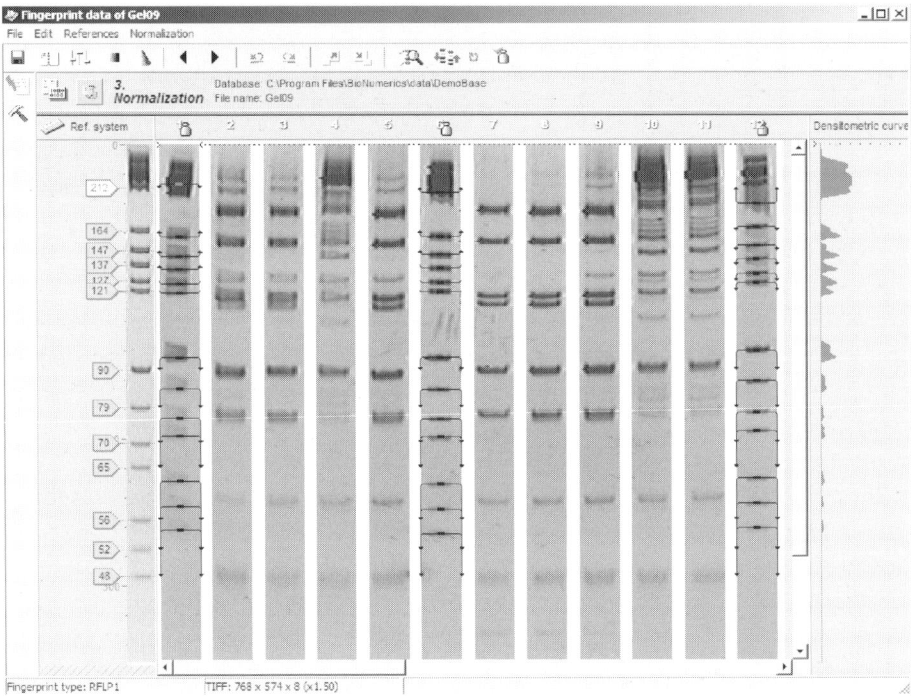

(Software used: The BioNumerics® Suite, version 4.0; Applied Maths B.V.B.A., Sint-Martens-Latem, Belgium).

Figure 3: The 'remapping' of two different reference systems (R01 and R02) by linking each position in the gel (run length expressed as a percentage) to a specific molecular weight or fragment length, allows the comparison of gels that have been run with different reference systems and are e.g. derived from different laboratories, or with large differences in time. In this conversion, some quality is lost due to the approximation of the conversion curves calculated.

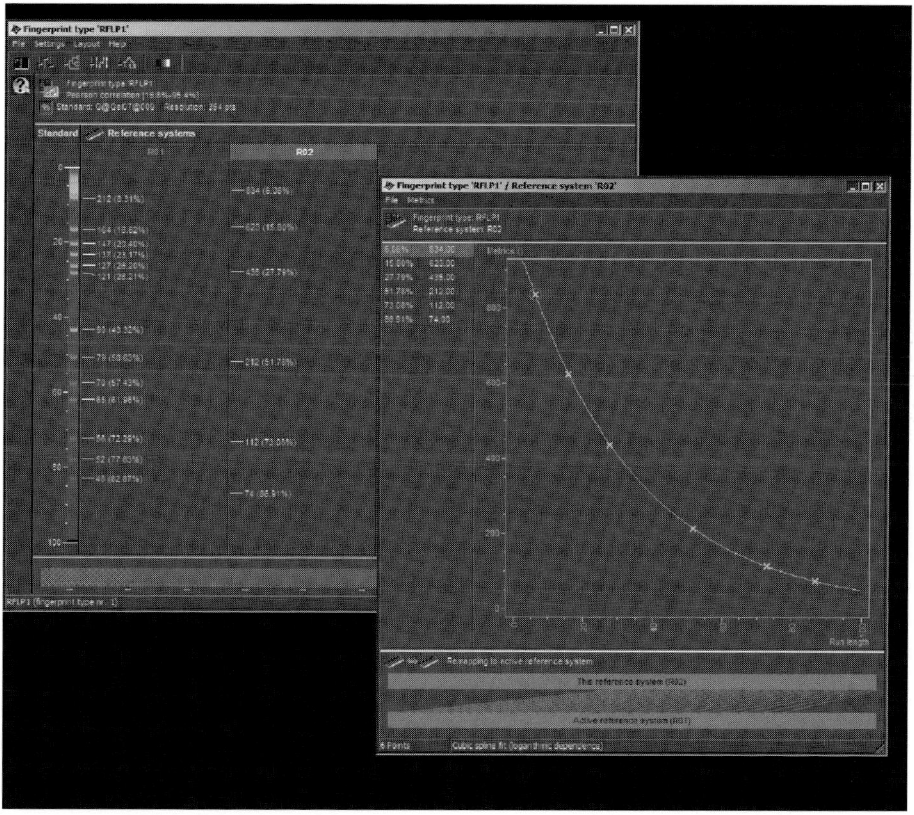

(Software used: The BioNumerics® Suite, version 4.0; Applied Maths B.V.B.A., Sint-Martens-Latem, Belgium)

Besides the quality of the data, quality of the analysis needs to be considered too. E.g. the inclusion of outliers should confirm the expected homogeneity of the dataset, and confirm the validity of the algorithms used. Validation techniques applied on dendrograms, especially when using band based similarity coefficients followed by UPGMA analysis, will show the reliability of any branching point (Figure 4). For sequences, the use of bootstrap algorithms is commonly used to estimate the reliability of a phylogenetic tree.

Gathering good quality data is expensive, but could bring about **commercial or strategic importance**. Therefore, it is to be expected that many data acquired will never become available to the public and will be kept in local, confidential databases. For security reasons, large research institutes (private or public) will not send precious sequences over the Internet for e.g. a BLAST analysis against public data. Preferentially they will try to store and manage the public information locally. As a consequence bioinformatics applications will need to deal with these large database requirements and provide proper potential to retrieve information in a fast but reliable way. When shared by many local users, these databases will also need to offer data protection with specific access priorities, which tend to make a functional implementation more difficult. In many cases only the above mentioned true relational database systems will be able to match all of these requirements.

Linking and combination of different data types brings new insight. Intelligent integration of data on database and analysis level is inevitable for data comparison and validation. Each data type, however, needs particular tools for optimal analysis: e.g. traditional alignment algorithms fail on chromosomal sequences, and traditional cluster analyses like UPGMA do not allow to mine a set of large micro-arrays due to time problems (time needed is a quadratic function of the number of entries).

On the other hand, the value of proteomics data e.g. will increase dramatically when links can be made to pathways or to DNA or amino acid sequence data, which on their turn can be linked to secondary or tertiary structures and to functionality.

Homologies between expression patterns obtained on the proteome and the genome level as well as gene expression profiles of genes linked to known biochemical pathways will fortify a conclusion. Therefore, the potential to link these different databases in an intelligent way will be very helpful in any data mining approach, but will necessitate a modular system that goes much further than the database level.

Figure 4: UPGMA dendrogram obtained from a similarity matrix, calculated with the Pearson Product Moment Correlation Coefficient for a set of RFLP patterns, generated for an artificial set of bacteria

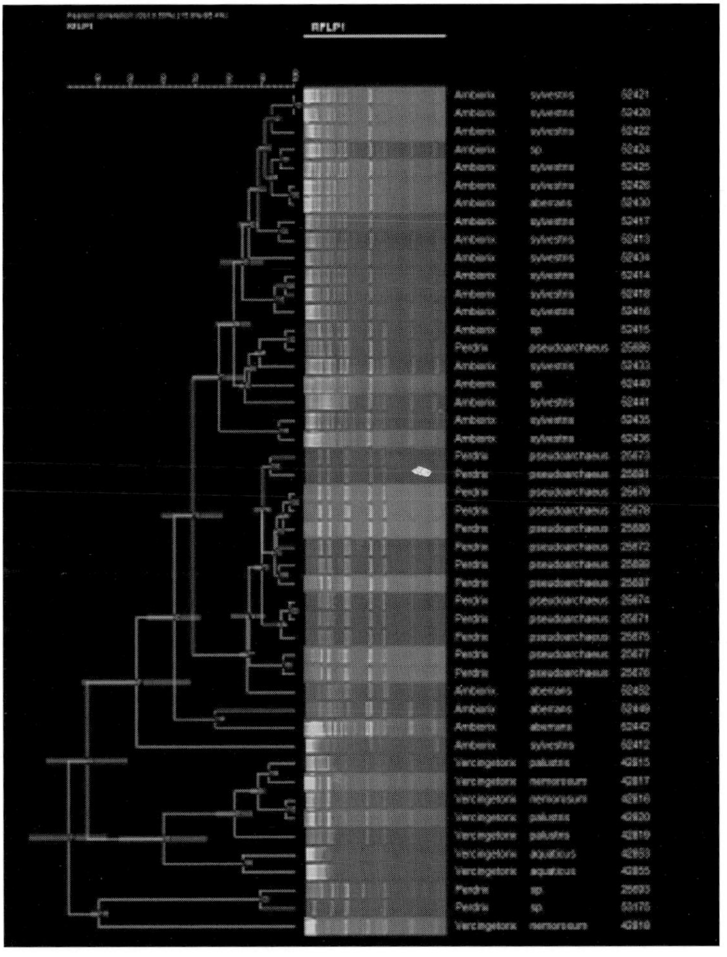

For each branching point the cophenetic correlation has been calculated, showing the reliability of the clustering for each individual branch (numbers between 0 and 100%). The gray bars indicated the 'error flags', representing the standard deviation for each clustering level, reflecting the degree of simplification resulting from the UPGMA approach. Optimization settings and zone used for comparison are indicated in the top line. *(Software used: The BioNumerics® Suite, version 4.0; Applied Maths B.V.B.A., Sint-Martens-Latem, Belgium)*

The use of an **adapted database system** will be a good start to help solve many of the above problems. Biological data can be organized in mainly three ways: (i) flat text files, (ii) object oriented databases, and (iii) relational databases.

Flat text files consist of text and consequently should be queried as such. The latter is very easy, is mostly platform independent and is therefore ubiquitous. However, searches in texts are always slow due to multiple file management operations. To speed up this process, files can possibly be indexed and structured (e.g. like EMBL or GenBank sequence formats), but complex, integrative queries will remain very complicated, if not impossible.

Object oriented databases organize data into a hierarchy of concepts or *classes*, each having a set of *attributes*, which can have typed values. In this way of data organization *classes* can *inherit* values of *attributes* from the parents in the hierarchy, giving rise to a major advantage of this type of databases, namely that they can represent extremely difficult relationships. However, queries are not as efficient as for the relational database model.

In a *relational database*, all data are represented in tables and pointers or links; the latter represent relations within the data. By adding tables and links, a relational database can be extended easily and consistently. Setting up a relational database is more complicated than a flat text file system and requires some preliminary consideration. However, once functional, commercial software will manage many options for maintenance and its use is extremely fast for storage, search and retrieval for even the most complicated query. Due to the structured nature of these databases, it is easy to hide or protect particular data for a certain category of users (e.g. local researchers versus external customers). A limited example of such a table structure for a diversity of biological data is given in Figure 5.

Setting up of such an integrative database is still depending on a number of conditions that need to be met, apart from all quality and compatibility hurdles mentioned above. These include, (i) the sharing of at least one common table or field to link new tables; (ii) the availability of at least one unique identifier to link new information to existing entries or trigger the creation of new entries; and (iii) the use of a uniform vocabulary (e.g. abbreviations of units, uniform accession codes or enzyme numbers, etc. Still, using a relational database system, integration of data becomes more feasible (taking into account all above hurdles!), and above all, allows extremely flexible querying.

Figure 5: Example of a table structure of a relational database, which describes the relations between the database entries, characterized by their genus name, species name, date and source of isolation (ENTRYTABLE), and the result of phenotypic tests such as antibiotic resistances (table Antibiotics), fatty acid analysis (FAME), fingerprint data (table FPRINT, with all technical parameters related to normalization, display band information, ...), and sequences (SEQUENCES). *(Software: The BioNumerics Suite, version 4.0; Applied Maths B.V.B.A., Sint-Martens-Latem, Belgium).*

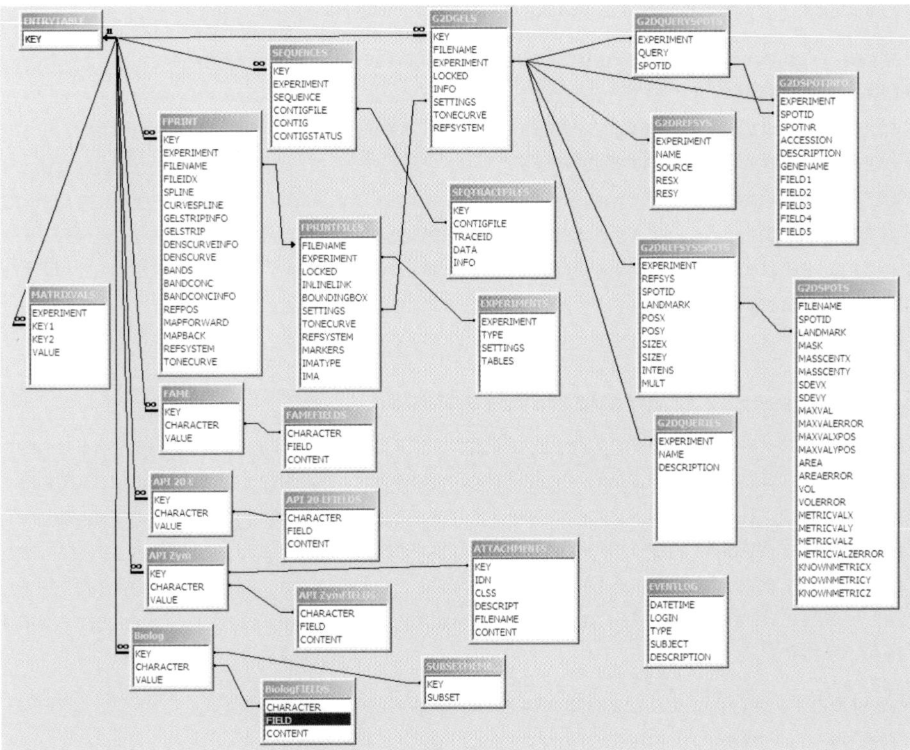

A final point of consideration is the exchange of information between databases. The transparent and structured flow of information between research institutes, between research institutes and governments or industries, or between governments and industry can considerably speed up the development of new products or improve the introduction of the latest technologies in e.g. the monitoring of quality or in certification.

As discussed above, the problem of standardization is not only related to data exchange protocols but variation related to small technological or protocol differences may have a much larger impact on the efficiency of data exchange. Still there are many cases where this exchange has been functioning extremely well. As an example we will discuss below the creation of a PFGE network with a central database of food-borne microorganisms *(Salmonella, Listeria, Escherichia coli)*, managed by the center of disease control (CDC) in Atlanta (the PulseNet initiative).

For certification reasons, electronic records needed to get the same value as (hand written) paper records. To that extend, the USA government has set forward standards for electronic data exchange with its services. One example of such standard is the *'Food and Drug Administration (FDA) Title 21 Code of Federal Regulations (CFR) Part 11. — Electronic Records; Electronic Signatures'.* With this regulation, the United States' FDA provides a standard for regulated companies for data security and integrity, electronic records, and electronic signatures. The regulation may be mandatory for the pharmaceutical and biotechnology industry because the regulation affects all drug and pharmaceutical products intended for the US market.

3. An example of an integrative solution

The BioNumerics® Platform (Applied Maths B.V.B.A. Sint-Martens-Latem, Belgium), used by the PulseNet initiative, described below is composed of a modular software that offers both a solution for many of the above mentioned data management problems, including both the database aspect as well as the data analysis and comparison aspect (Figure 6). This modular system can be expanded over time, can easily be adapted to new functionalities and will assure long-term data management.

The solution can handle almost any data type (fingerprint data (e.g. PFGE, DGGE, mass spec profiles, fragment analysis profiles derived from sequencers, band tables, etc.), sequences (NA or AA, including full genomes), micro-array data, character data, matrix data, 2D proteomics data, etc...) retrieved from e.g. text files, TIFF images, Microsoft® Excel files, etc... and stored in a local, pseudo-relational database managed by the software itself or in an external multi-user relational database (Microsoft® SQL Server™, Oracle®, Microsoft® Access, PostgreSQL, MySQL, etc.) addressed directly via ODBC. The available toolbox allows the normalization of data (e.g. of fingerprint data, fragment length tables, or 2D protein gels) and resolves the data access problem through advanced scripting potential, allowing import and export of (almost) any non-encrypted data format.

Data comparison algorithms are optimized for each data type and size being analyzed. Also, proper tools are provided for data verification and for the evaluation of the quality of the groups or the results obtained (e.g. cophenetic correlation, bootstrapping, cluster violation, K-means partitioning, etc.). Further validation is achieved by the use of different algorithms, comparing alternative methods for grouping (cluster analysis versus dimensioning techniques such as Principal Component Analysis, Discriminant Analysis, Multi Dimensional Scaling, or special approaches such as Self-Organising Maps, MANOVA, Neural Networks or Support Vector Machines).

Figure 6: Modular structure of a data management and analysis package for biological data. All modules communicate with an integrated central database and can exchange information seamlessly. A uniform interface applies to all.

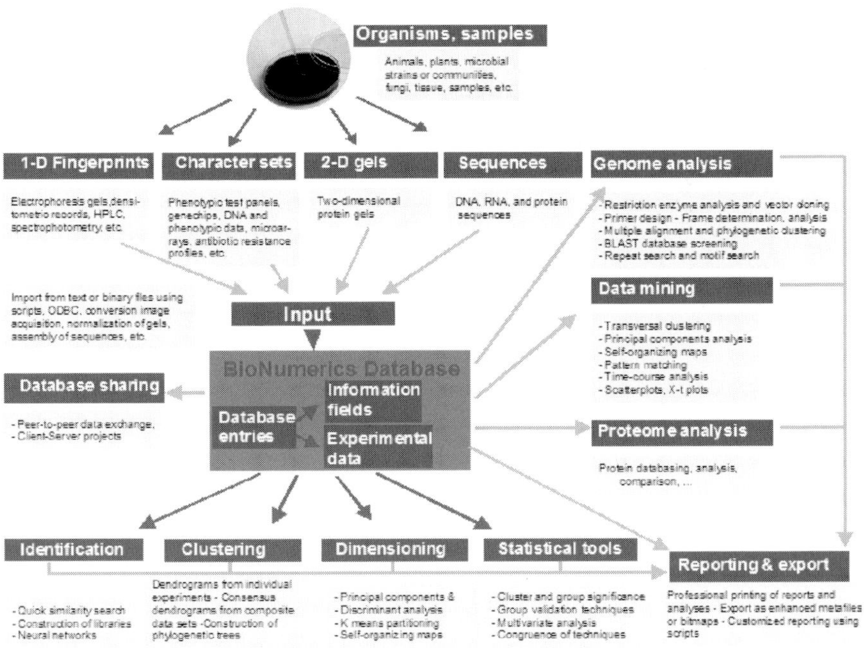

It is also possibility to make combined datasets, linking e.g. fingerprints to sequence- or character data to improve or fine-tune groups or identifications (see Composit dataset in Figure 1).

The most performing database setup is the option to use a fully relational database system, managed by e.g. Microsoft® SQL Server™, Oracle®, Microsoft® Access, PostgreSQL, MySQL, etc... Through a permanent ODBC link, the software can interact directly with the database tables or, if necessary, with a database view that represents only the data needed by the software. The interface is therefore fully depending on the software and offers standard user-friendly representations for e.g. gels, micro-arrays, or phenotypic tests.

For more advanced data mining the central relational database will allow complex querying with e.g. the linking of 2D proteomics data with the result of a micro-array expression experiment or with genomic information. The purpose can be to detect common patterns and confirm the validity of a selected set of genes / proteins, contributing to an advanced quality control, as discussed above.

The software is equipped with a database sharing tools module, taking advantage of readily available e-infrastructure to face the communication problems of today's life sciences and enabling sharing of knowledge and collaboration between scientists, between scientists and industry, and between researchers and the government. In combination with database solutions offered by e.g. Oracle®, compliancy with Title 21 CFR Part 11 can be guaranteed.

Finally, for more performing networks, like the example discussed below, the BioNumerics® software can be coupled to a Server solution (BNServer) which allows automatic services for data uploading, data downloading, identification, alarm generation, etc... Client / server communication protocols are based on XML or HTML, and can be organized through direct TCP/IP connectivity or VPN.

The setting up of a user-defined network is facilitated through the scriptability allowing user-defined login screens and protocols and offering different levels of data protection (see below for an example).

4. The PulseNet example

PulseNet (http://www.cdc.gov/pulsenet/what_is.htm) is the National molecular subtyping network for foodborne pathogens, which uses mainly Pulsed-field Gel

Electrophoresis (PFGE) for the subtyping of pathogens, along with some other information as serotyping, antibiotic resistances, etc... After a large outbreak of foodborne illness in 1993, caused by the bacterium *Escherichia coli* O157:H7 in the western United States, CDC set-up PulseNet in 1996 in conjunction with the state public health laboratories in Massachusetts, Minnesota, Texas, and Washington. Since then, the network has expanded to 7 *Area Laboratories* which included Massachusetts, Michigan, Minnesota, Texas, Utah, Virginia and Washington, 43 *State Public Health Laboratories* and 5 *City/County Laboratories* have come on line. PulseNet Canada, a similar initiative in Canada, has six *Provincial Public Health Laboratories* and a *Federal Food Safety Laboratory* that joined the network in 1999.

PulseNet is an early warning system for outbreaks of foodborne disease *(Escherichia coli*, *Salmonella*, *Shigella*, *Campylobacter* or *Listeria)*. The network identifies and labels each "fingerprint" pattern and permits rapid comparison of these patterns through an electronic database at the Centers for Disease Control and Prevention (CDC) to identify related strains. (http://ftp.cdc.gov/pub/infectious_diseases/iceid/2002/pdf/hunter.pdf) PulseNet's success lies in its ability to recognize widely dispersed clusters of foodborne illness that may indicate that an outbreak is occurring. Early detection of a cluster is achieved through active surveillance, which in the case of PulseNet means performing routine PFGE subtyping of routine isolates received in each of the participating laboratories and submitting the DNA fingerprints of these isolates to the national database.

At individual State Health Laboratoires in the United States, the PFGE patterns are gathered according to well-defined protocols and the digitized patterns are normalized and analyzed using the BioNumerics® software. BioNumerics® can submit these patterns to the National database located at CDC in Atlanta, GA. This National Pattern Database collects and analyses all incoming patterns with the purpose to alert PulseNet participants of a possible multistate outbreak so that a timely investigation can be done. Since PulseNet was established in 1996, there have been many instances in which real-time PFGE subtyping led to the identification of outbreaks, nationally and internationally, even when there was no initial epidemiological evidence suggesting that outbreaks were occurring.
(http://www.aphl.org/Infectious_diseases/idconfdocuments/molecular/besser.pdf)
(http://ftp.cdc/gov/pub/infectious_diseases/iceid/2002/pdf)

The success of the PulseNet project has stimulated other initiatives in the field.

Examples include:

1. The Foodborne and Diarrheal Diseases Branch at CDC in collaboration with the Division of Vector Borne Infectious Diseases, Bacterial Zoonotic Diseases Branch (BZB), NCID/CDC is expanding PulseNet to include a Bioterrorism (BT) module starting with two Category A bioterrorism bacterial zoonotic agents, *Yersinia pestis* and *Francisella tularensis,*

   (http://www.cdc.gov/pulsenet/newsletter/spring_2003.htm#7).

   Dr. Kristy Kubota is leading the development of the PFGE protocols and BioNumerics® client database platform. The plan is to adopt the developed subtyping protocols for use in the Laboratory Response Network (LRN). Most, if not all the LRN laboratories, already have the capability of performing PFGE and are using BioNumerics® PulseNet client databases to connect to the national PulseNet server to upload and compare patterns. The current BioNumerics® client scripts already established for the foodborne pathogens need minor modification to be used for these pathogens in the event of a bioterrorism attack as well as for routine surveillance.

2. Giving the success of the typing approach in early warning of food-borne disease detection, the CDC is currently exploring the possibility to expand the current network to include (large) food producing companies. The required food quality monitoring protocols implemented at the production sites, could be adapted to include PFGE of possible contaminants. By submitting these patterns, any producer could easily find out if e.g. raw materials bought can be considered under risk or not. The current experience, both in respect to reproducibility and uniformity of data, has confirmed the feasibility of this approach.

3. In November, 2002, Dr. Lai-King Ng of the National Laboratory for Enteric Pathogens, the National Microbiology Laboratory, Population and Public Health Branch announced that PulseNet Canada, in a next phase will also include MRSA and VRE PFGE typing to the electronic surveillance system. This will clearly expand the scope of the project to a more clinical setting, widening the epidemiological reach of the network.

4. Other PulseNet projects worldwide:

   On December 12-13, 2002, a meeting was held in Honolulu, Hawaii to explore the possibility of setting up PulseNet Pacific Rim. Representatives from CDC,

APEC (Asia Pacific Economic Cooperation), WHO, EnterNet, and the Washington State Department of Public Health and 14 representatives from 12 Pacific Rim countries/areas agreed that a network for food-borne surveillance in the region should be formed. It was generally concluded that this network for which the name *PulseNet Asia Pacific* was proposed, would benefit by functioning as a part of the global PulseNet network. A first BNserver was set up by Dr. Brent J. Gilpin (Institute of Environmental Science & Research Limited, Christchurch Science Centre, and Christchurch, New Zealand).

A consortium of European scientists headed by Dr. Peter Gerner-Smidt of the Statens Serum Institute, Copenhagen, Denmark (formerly Visiting Scientist, PulseNet Europe, Centers for Disease Control and Prevention, Atlanta, GA), are working towards the establishment of a PulseNet Europe. In Denmark, Statens Serum Institute and the Danish Veterinary Laboratory collaborate jointly in a project that had the aim of improving the background for tracing of infection sources for human salmonellosis (http://www.rivm.nl/bibliotheek/rapporten/284500017.pdf). The project includes a harmonization and standardization of methodology applied in the two laboratories using PFGE and amplified fragment length polymorphism (AFLP). The purpose is to compare results and evaluate feasibility of exchange in electronic form via the Internet. Build on this experience, the extended consortium recently succeeded in obtaining European financing to further elaborate this initiative on a European scale. The BioNumerics® Platform linked to BNServer was proposed for the underlying bioinformatics structure.

5. Other projects worldwide:

In 2000 a European Union research “New generation genetic markers and techniques for the epidemiology and control of tuberculosis” (Q2K2-CT-2000-630) was started (Concerted Action on TuBerculosis, CA on TB, RIVM, Utrecht, The Netherlands, https://hypocrates.rivm.nl/bnwww/index.html). The overall objective of this project was not only to reinforce the collaboration between laboratories, but also to explore the possibilities of inter-laboratory molecular epidemiology as a tool to monitor and possibly understand transmission of tuberculosis. In order to optimize data exchange (DNA fingerprint analysis) between associated reference laboratories, the BioNumerics® software was used. A better understanding of the epidemiology of tuberculosis, and the resistant *Mycobacterium tuberculosis* isolates in particular, will contribute significantly to its effective control; the wider the network of participating laboratories, the more interesting the resulting databases that can be constructed.

5. Conclusions

Modern bioinformatics is a complex happening that needs to deal with increasing amounts of data, with quality issues as well as with efficient data mining and analysis. High flexibility is required to cope with an ever-changing diversity of data and maintain exchange potential at any time. Although complex professional databases are required, querying should be simple, flexible and effective.

The BioNumerics® Platform is composed of a modular software suit that approaches solutions to many of these problems. By using the BNServer application, automatic response routines can be set up at the server level, providing a 24hours / day tool for e.g. quality monitoring or early warning systems.

The platform should be continuously extended to cope with new emerging disciplines and techniques. The rapid growth of databases with pathway information, SNP information, and increasing numbers of algorithms, require a continued development and optimization. Modular approaches can guarantee a continued functionality in this respect.

Its successful use over several years in a number of ongoing research and quality monitoring networks has proven that bioinformatics in a wider network is feasible and can be of use in improving and speeding up research efforts in many fields of biology.

References

1. Vandamme, P., Pot, B., Gillis, M., De Vos, P., Kersters, K. and Swings, J. (1996). Polyphasic taxonomy, a consensus approach to bacterial systematics. *Microbiological Reviews*, **60**: 407-438.

Where is the Genome Sequence leading us? A UK approach to the functional genomics of *Streptomyces coelicolor*

K. F Chater[1], P. Dyson [2] and C. P Smith [3]

[1] *John Innes Centre, Norwich Research Park, Colney, Norwich NR4 7UH, UK*

[2] *Department of Genetics, University of Wales Swansea, Swansea SA2 8PP, UK*

[3] *Department of Functional Genomics, University of Surrey, Guildford GU2 7XH, UK*

1. Introduction

Since the 1940s when Selman Waksman first demonstrated that they produced valuable antibiotics, streptomycetes have been the pre-eminent source of these vital elements of modern medicine. In choosing to study *Streptomyces* genetics in 1954, the young David Hopwood was very conscious of this aspect of their biology. By a happy chance he focused on *S. coelicolor* A3(2), in part because of the possible usefulness of its blue pigment as a genetic marker. It was not until the early '70s, when he found out that the blue pigment was a polyketide antibiotic, actinorhodin, that he went on to use it to understand the mechanisms of polyketide biosynthesis and to provide the first examples of combinatorial biosynthesis: the process in which genes from different biochemical pathways are combined to permit the host to produce new secondary metabolites. Thus, *S. coelicolor* began to approach the orbit of industrial microbiology more closely.

The sequencing of the *S. coelicolor* genome was carried out at the Sanger Centre in the UK, under the guidance of David Hopwood about 100 km away at the John Innes Centre. It was funded by the Biotechnology and Biological Research Council (BBSRC), backed up by the Beowulf Initiative of the Wellcome Trust, and was achieved by the sequencing of an ordered set of 325 overlapping cosmid and BAC inserts. This approach has provided us with a wonderful resource (i.e., a set of completely sequenced clones) that has been very important in functional genomics.

When the genome sequence of this representative streptomycete was completed in 2002, we had our first glimpse of the whole organism from the inside [1]. It rapidly became clear that the sequence was going to be a constant reference point for everyone wanting to use gene-based approaches to the industrial exploitation of this genus. Many new insights into physiology and metabolism were quickly gained, and there are many more to come. The chromosome was confirmed to be linear and, at about 8.6 Mb, to be twice as big as the circular *E. coli* K-12 chromosome. It was deduced to contain 7,825 genes – more than were annotated in the genome of the eukaryote *Saccharomyces cerevisiae.* There is an exceptionally high proportion (10%) of likely regulatory genes, including those for about 65 sigma factors, 84 sensory histidine protein kinases, 80 response regulators, and 34 serine-threonine protein kinases [1, 13] (M. Hutchings, P Hoskisson and G. Chandra, *pers. comm.*). Putting these together with the nearly 1000 genes devoted to transport, a picture emerges of an organism that is highly sensitive to, and interactive with, its environment. Of importance to future metabolic engineering, there are many examples of apparent multiple isoforms of metabolic enzymes. Particularly strikingly, the number of secondary metabolites potentially made by *S. coelicolor* was revealed to be several times as many as the six previously known, with about 5% of the genome being devoted to secondary metabolism.

While the sequence was being compiled and annotated, the Sanger team established a database (ScoDB) straightforwardly derived from AceDB, which had been designed initially for the genome of the nematode *Coenorhabditis elegans.* Although AceDB was invaluable as a stopgap, it suffered from a number of limitations, including ease and speed of access, built-in impediments to updating, and significant eukaryotic baggage. In addition, it was not similar in its user interface to the emerging standards being applied to other genomes. Because of this, bioinformaticist Govind Chandra rebuilt the database from scratch, to give ScoDB2. He curates this database (streptomyces.org.uk) at the John Innes Centre where he is embedded in the large *Streptomyces* research group. This biological context maximises the extent to which the database can be enhanced for the needs of the user community. At the time of writing, gene annotation (in some cases including preliminary gene ontology) can be accessed by: gene name (e.g. *glgB*); gene number (e.g. SCO4228); position in the genome, either using numerical coordinates or a graphical representation that is part of the display; or text search. The display also shows the positions of transposon insertions available in cosmids, which are usable for the rapid disruption of the gene in question (see below).

As the sequencing began, the BBSRC encouraged the UK *Streptomyces* community to apply for further funds, in order to establish effective microarray and proteomic approaches. Even though we hardly knew what these terms meant, a consortium was formed and funding secured. Work started in 1999. A year later, further funding was won to allow the development of genome-based high throughput methods for gene disruption. In the rest of this article we outline how far we have come in the intervening years.

2. Bringing *Streptomyces* DNA microarrays on-stream: the first data begin to flow

The availability of the *S. coelicolor* genome sequence has allowed the design and production of comprehensive DNA microarrays representing most of the annotated ORFs (Fig 1). This has been achieved both by us and by the groups of Stanley Cohen and Camilla Kao in Stanford University; see Huang *et al.,* [12]. Both efforts have initially focussed on producing PCR-based probes for each ORF. The care we have taken to design primers that avoid production of PCR products that might cross-hybridise with other sequences in the genome is described at http://www.surrey.ac.uk/SBMS/Fgenomics/Microarrays/. This is particularly important for *S. coelicolor* because of the significant numbers of gene families; thus, where possible, small microarray probes were produced to avoid cross-hybridisation between paralogues. Our current PCR-based array contains 7,333 probes (transposons and IS elements have been excluded). More recently, in conjunction with MWG Biotech, we have designed and produced a 50-mer oligo set that represents all unique ORFs (currently 8,006) of *S. coelicolor*. Both array formats include ORFs from the 365-kb linear plasmid SCP1 [2]. To date the *Streptomyces* microarrays have been used mainly for three types of studies: (i) comparison of the genome composition of different streptomycete species and different strains of *S. coelicolor* A3(2) – a rapid and powerful application of arrays for tracing strain lineage and taxonomic relationships; (ii) comparisons of transcriptomes in a wild-type versus a particular gene knockout mutant – a rapid route to identifying genes directly or indirectly influenced by the particular gene inactivated in the mutant; (iii) analysis of gene expression at different stages of growth of a culture, or following a particular treatment – the identification of genes that follow a similar expression profile provides a powerful method of grouping them into common regulatory networks. An example of the latter is given in Figure 2, which illustrates a set of genes that are induced around the transition phase of growth on complex solid media.

Our global array-based studies of gene expression indicate that expression of up to half the total complement of genes is detectable during active growth on complex media. It has also been revealed that gene expression at the transition phase that precedes stationary phase can be dramatically altered following the deletion of single regulatory genes. This observation suggests that the disruption of single metabolic pathways can perturb large numbers of pathways, presumably because metabolic and regulatory networks are interconnected. More recently, efforts have been initiated to correlate microarray data with proteome and metabolite profile data from the same cultures. These integrative 'omic' approaches should eventually allow us to predict the phenotypic and metabolic outcomes of gene-knockouts or other artificial perturbations.

Figure 1: *Streptomyces coelicolor* PCR-based microarray produced by the UK consortium

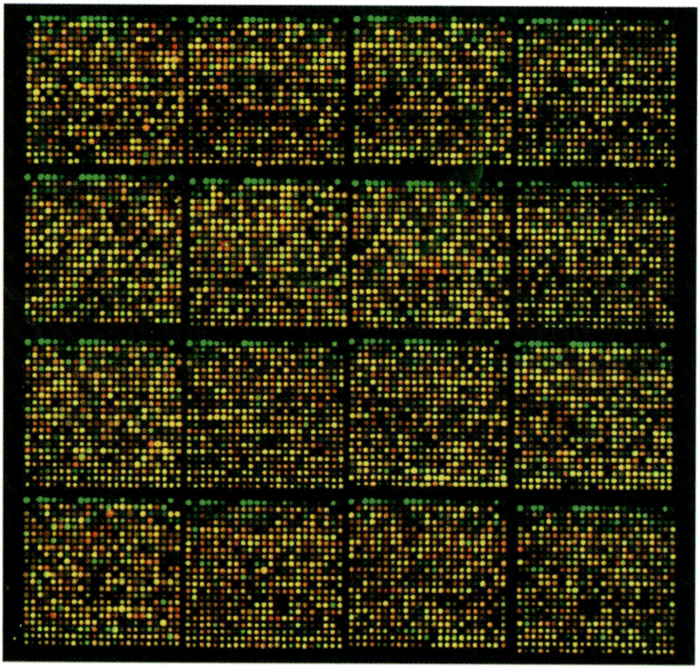

S. coelicolor total RNA was used to generate Cy3-labelled (green) cDNA and was co-hybridised with Cy5-labelled (red) genomic DNA.

The microarrays have now been distributed to more than 20 research groups worldwide. The first papers in which they have been used are now in preparation: for example, Eriko Takano and co-workers (manuscript submitted) have used the arrays to home in on targets for regulation by the SCB1 gamma-butyrolactone signalling molecule of *S. coelicolor.* If these arrays are taken up broadly by the research community, and this is coupled with adoption of common protocols, the opportunities for making meaningful comparisons of array data generated in different laboratories will be maximised. The PCR-based microarrays are now available to all laboratories (on a cost-recovery basis) and the new oligo-based microarray will follow shortly. Arrays of either format can be requested via http://www.surrey.ac.uk/SBMS/Fgenomics/Microarrays/html/Request.html.

3. Complexity of the proteome

After a substantial period of technical optimisation, not least to be able to avoid protein degradation by the proteases produced in abundance by streptomycetes, Andy Hesketh at the John Innes Centre was able to identify nearly 1000 of the protein spots separated by 2D polyacrylamide gel electrophoresis, by mass spectrometric analysis of tryptic digests (http://dbk.ch.umist.ac.uk/StreptoBASE/s_coeli/referencegel/) [10]. These spots represented the products of 770 genes. Thus, about 20% of genes give rise to more than one spot. Further analyses of selected examples of such multiple spots have shown that some of them are due to covalent modification. An example of this is PII, the product of the *glnK* gene, and an important regulator of cellular responses to changes in nitrogen availability [9]. In all other systems characterised to date, the properties of PII protein are modulated by its uridylylation, but in *S. coelicolor* it is adenylylated instead. The modified form is also subject to a specific proteolytic cleavage to generate a third spot. Specific proteolysis is one of the commonest causes of additional spots for particular gene products.

Proteomics is telling us some unexpected things about natural product biosynthesis. Under the growth conditions employed, protein spots involved in 8 of the secondary metabolic pathways recognised in the genome annotation were detected, including some for unknown end-products (thus revealing that these culture conditions were suitable ones in which to try to identify these end-products). In several of the pathways, but most notably in that for actinorhodin biosynthesis [11], there were examples of multiple spots, with cases of both covalent modification and proteolytic cleavage (Fig. 3).

Figure 2: Identification of a subset of genes up-regulated at the 'transition phase'

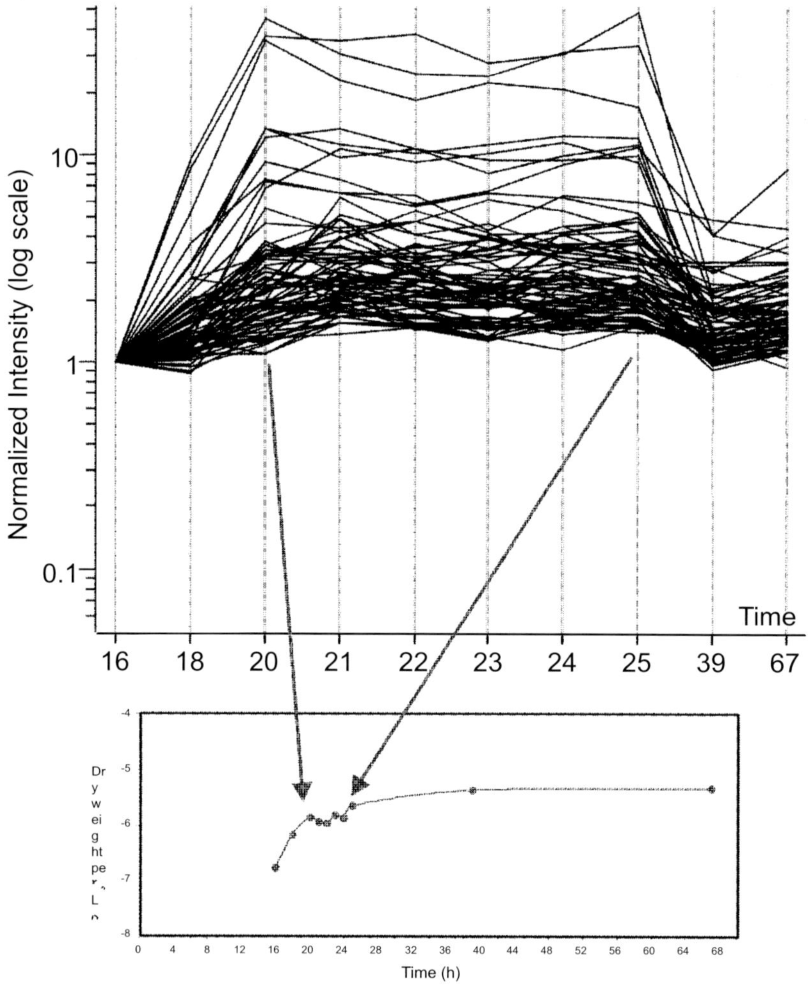

Wild-type *S. coelicolor* was cultivated on cellophane placed on agar medium and harvested at ten time points. The gene expression data were normalized relative to the first time point. Many of the co-expressed genes in this graph correspond to known and hypothetical genes involved in the regulation and production of secondary metabolites.

The significance of these for the activity of the relevant pathways remains to be investigated, but clearly the use of enzymes expressed in *E. coli* may mask some aspects of biosynthetic enzyme action. Another surprise was the discovery that one enzyme of actinorhodin biosynthesis is found almost exclusively outside the cytoplasm, as a protein that can be washed from cell walls. Does this mean that an actinorhodin precursor is secreted, acted on by the enzyme, and then re-imported for further biosynthetic steps?

Various technical limitations of the approach used for our proteomic analysis mean that many proteins cannot be seen, even if they are produced in significant amounts. The limitations include size (very small or very large proteins are missed) and location in membranes. Taking this into account, some simple extrapolations indicate that about half the theoretical proteome is represented at levels potentially detectable by silver staining in the cultures we are typically studying. This is in good agreement with the microarray analysis of global transcription activity mentioned above.

4. Zapping the genome: the application of efficient *in vitro* transposon mutagenesis

Every genome-wide analytical process gives rise to the need for genetic analysis – in the simplest case, the need to knock out genes and evaluate the phenotypic consequences. Thus, genome annotation may suggest that certain genes are involved in an interesting aspect of cellular physiology, or changes in the levels of transcripts or proteins discovered in microarray or proteomic analysis of a regulatory mutant may tentatively implicate them in a common process. To complement genome-wide microarray and proteome analysis, mutation of all 7,825 genes is a desirable goal (albeit a daunting one, based on conventional approaches for creating targeted gene knock-outs in *Streptomyces*). One route for constructing an extensive mutant library is to employ *in vivo* transposon mutagenesis. Several transposable elements have been employed with varying success for this purpose in *Streptomyces*. Indeed, through the efforts of Kay Fowler and Tobias Kieser, at the John Innes Centre, extensive mutant libraries of *S. coelicolor* have been constructed recently, using Tn*4560* delivered on an unstable plasmid [8].

ORFs are overlooked and, with a view to achieving saturation mutagenesis, it is perceived that knocking these out using PCR targeting (see below) is likely to be more economical than sequencing a greater number of mutated cosmids. Importantly, for curation and distribution of the mutant libraries, individual insertions are stored as mutant cosmid DNA, as *E. coli* clones and, after intergeneric conjugation and selection for allele replacement, as mutant spore suspensions.

The versatility of the technology has been demonstrated with genetic dissection of genes located on cosmid SC3C3 involved in adaptation to conditions of high external osmolarity [3]. Two adjacent genes, *osaAB*, encoding a hybrid histidine kinase and a response regulator, were found to play an integral role in signal transduction enabling osmoadaptation. Moroever, their disruption also resulted in a five-fold increase in antibiotic production under hyperosmotic growth conditions. Although both genes are upregulated in conditions of high external osmolarity, as shown by analysis of *egfp* expression in Tn*5062* insertional mutants, transcriptional analysis showed that *osaB* is regulated independently from *osaA*; this is unusual for a histidine kinase - response regulator gene pair. Further analysis exploited another versatile aspect of shuttle mutagenesis: mutations can be transferred very readily to a variety of different genetic backgrounds, an important aspect for analysing genetic interactions, and even to other related species such as *S. lividans*.

To facilitate construction of strains containing more than one gene disruption, exchange cassettes have been engineered to permit the substitution of the apramycin resistance determinant in an existing Tn*5062* insertion by another resistance marker, for example that for hygromycin. Manipulation of Tn*5062* insertions to swap resistance markers is made possible by exploiting PCR targeting (see below). Bertolt Gust at the John Innes Centre has also used this procedure to create in-frame deletions of genes containing a Tn*5062* insertion, a procedure that relies on the presence of a rare *Afl*II restriction site in the transposon. These examples demonstrate the versatility of the mutant cosmid library; as a consequence, the worldwide *Streptomyces* community can now interrogate the complexity of the streptomycete genome and address the implications stemming from analysis of large microarray and proteome data sets.

4.1 Customised changes to the *Streptomyces* genome

High throughput transposon mutagenesis provides access to knockout mutations in a significant fraction of *S. coelicolor* genes, but in order to fill in any gaps it is

Figure 4: Organisation of Tn5062. The transposon consists of two inverted repeats (ME; Mosaic Ends) flanking an internal region containing the following elements: (1) a sequence consisting of translational stop codons in all three reading frames (stop), (2) a consensus streptomycete ribosome site (RBS), (3) a promoter-less copy of the green fluorescent protein gene (*egfp*), (4) an apramycin-resistance gene (*aac3(IV)*) flanked itself by two T4 transcriptional terminators (T_4), and (5) the RK2 origin of transfer (*oriT*). Also indicated are positions of two sequencing primers, EZL2 and EZR1, used to precisely determine the location of each insertion.

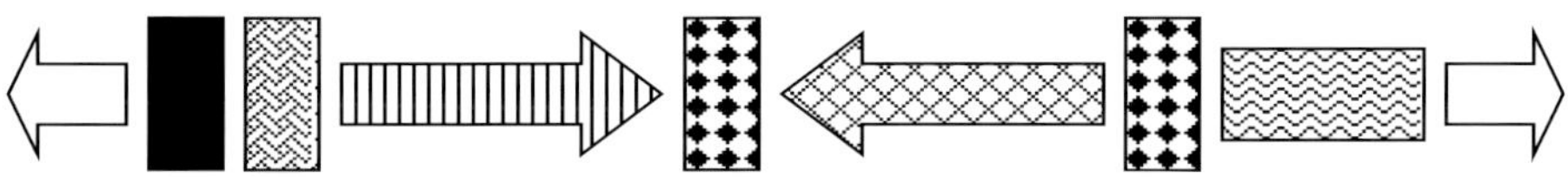

useful to have an efficient system of targeted mutagenesis. Such a resource has been developed by Bertolt Gust [8] at the John Innes Centre, starting from the PCR targeting system described by Datsenko and Wanner [4]. In a nutshell, the procedure consists of three steps. First, a resistance gene selectable in both *E. coli* and *Streptomyces* is amplified by PCR, using primers that have 39 bp 5' extensions corresponding in sequence to the boundaries of the desired gene replacement (Fig. 5). Second, the amplified DNA is used to transform a hyper-recombinogenic *E. coli* strain harbouring a cosmid clone carrying the gene to be disrupted, resulting in replacement of the target gene by the PCR-amplified fragment. Thirdly, the cosmid containing the disrupted gene is introduced into *Streptomyces*, where there is usually a high frequency of double crossover replacement of the target gene. The last step is made easier by the inclusion in the resistance gene cassette of the transfer origin of a wide host range plasmid, which makes it easy to transfer the altered cosmid into *Streptomyces* by conjugation, just as indicated above for introducing Tn*5062*-

disrupted genes into *Streptomyces*. Using this technique, about 200 mutants have been constructed so far.

This basic version of the technique has many variants, which make it straightforward to introduce all sorts of genetic changes into the *Streptomyces* genome. To date, these include unmarked in-frame deletions; point mutations; promoter swaps; translational fusions to the GFP reporter gene; and the introduction of immune tags. Because the changes are all made at the correctly located chromosomal position, there is minimal disturbance to the normal regulation of the target gene (unless it is actively sought!).

PCR targeting is especially useful for the analysis of complex gene clusters, in which it is usually desirable to avoid mutations that are potentially polar (i.e. affect the expression of genes downstream in an operon). It is therefore particularly valuable for the study and manipulation of the large gene clusters that are involved in antibiotic biosynthesis (e.g., Eustáquio *et al*., [5]).

Figure 5: Typical cassette used for PCR targeting. The components of the cassette are as follows: P1, P2, 20 bp sequences used as templates for amplification by PCR; FRT, directly repeated target sites for the yeast site-specific FLP recombinase, used to excise the intervening DNA to generate non-polar deletions; *oriT*, wide host-range transfer origin, permitting mobilisation of plasmids between *E. coli* and *Streptomyces*; *aac(3)IV,* apramycin resistance determinant selectable in both *E. coli* and *Streptomyces*; arrows, primers for PCR amplification, with 39-bp 5' ends homologous to the gene to be targeted and 3' ends homologous to P1 and P2 [8].

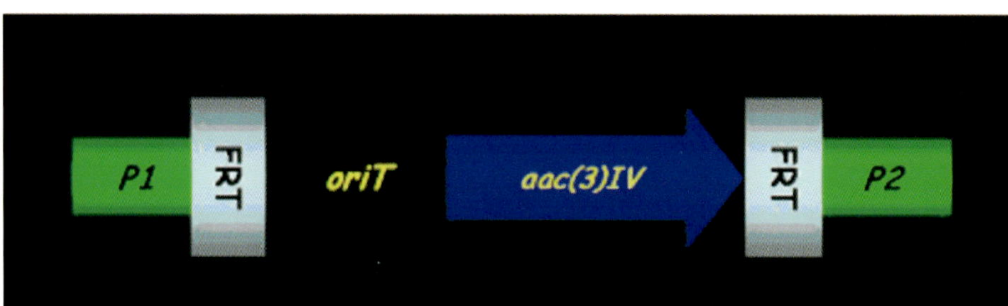

5. What next?

The rapid technical advances described here are beginning to generate previously unimaginable amounts of data. This by itself provides a new challenge – how can it all be managed and integrated? A major current preoccupation is to develop the bioinformatic tools that will permit this, and allow user scientists to access the resources in a user-friendly manner. The systems being developed, at the University of Manchester by Andy Brass, Norman Paton and Mike Cornell, and at the University of Manchester Institute of Science and Technology by Douglas Kell and Giles Velarde, will, in future, be the seed of a growing on-line picture, to which will be added such information as metabolomics, protein structures, detailed phenotypes of mutants, and spatial and temporal information about cellular organisation. Such a comprehensive "world-view" of these remarkable organisms will provide an astonishingly rich context in which to engineer secondary metabolism for the benefit of us all.

Acknowledgements

The work described here was mainly funded by grants from the BBSRC (IGF12432 and FGT11408 to KFC, IGF12431 to PJD, and FGT11407 to CPS). We thank our colleagues Govind Chandra, Andy Hesketh, Bertolt Gust, Paul Herron, Amy Bishop, Graham Hotchkiss, Vassilis Mersinias and Giselda Bucca for carrying out the experiments on which this article is based.

References

1. Bentley, S.D., Chater, K.F., Cerdeno-Tarrega, A-M., Challis, G.L., Thompson, N.R., and James, K.D. (2002). Complete genome sequence of the model actinomycete *Streptomyces coelicolor* A3(2). *Nature,* **417:** 141-147.
2. Bentley, S.D., Brown, S., Murphy, L.D., Harris, D.E., Quail and M.A., Parkhill, J. (2004). SCP1, a 356,023 bp linear plasmid adpted to the ecology and developmental biology of its host, *Streptomyces coelicolor* A3(2). *Molecular Microbiology,* **51:** 1615-1628.
3. Bishop, A., Fielding, S., Dyson, P. and Herron, P. (2004). Systematic insertional mutagenesis of a streptomycete genome: a link between osmoadaptation and antibiotic production. *Genome Research,* **14**: 893-900.
4. Datsenko, K.A. and Wanner, B.L. (2000). One-step inactivation of chromosomal genes in *Escherichia coli* K-12 using PCR products. *Proceedings of the National Academy of Science USA*, **97**: 6640-5.
5. Eustáquio, A.S., Gust, B., Luft, T., Li, S.M., Chater, K.F., and Heide, L. (2003). Clorobiocin biosynthesis in *Streptomyces*. Identification of the halogenase and generation of structural analogs. *Chemistry & Biology*, **10:** 279-88.

6. Gehring, A.M., Nodwell, J.R., Beverley, S.M., and Losick, R. (2000). Genomewide insertional mutagenesis in *Streptomyces coelicolor* reveals additional genes involved in morphological differentiation. *Proceedings of the National Academy of Science USA,* **97**: 9642-7.

7. Goryshin, I.Y. and Reznikoff, W.S. (1998). Tn*5 in vitro* transposition. *Journal of Biological Chemistry,* **273**: 7367-7374.

8. Gust, B., Challis, G.L., Fowler, K., Kieser, T. and Chater, K.F. (2003. PCR-targeted *Streptomyces* gene replacement identifies a protein domain needed for biosynthesis of the sesquiterpene soil odor geosmin. *Proceedings of the National Academy of Science USA*, **100:** 1541-1546.

9. Hesketh, A., Fink, D., Gust, B., Rexer, H.-U., Scheel, B., and Chater, K. (2002a). The PII nitrogen regulatory protein GlnK in *Streptomyces coelicolor* A3(2) is modified by adenylylation rather than uridylylation, and the GlnD and GlnK homologues are functionally dissimilar to their nitrogen regulatory counterparts. *Molecular Microbiology,* **46:** 319-30.

10. Hesketh, A., Chandra, G., Shaw, AD., Rowland, J.J., Kell, D.B., Bibb, M.J. and Chater, K.F. (2002b). Primary and secondary metabolism, and post-translational modifications, as portrated by proteomic analysis of *Streptomyces coelicolor. Molecular Microbiology,* **46**: 917-32.

11. Hesketh, A. and Chater, K.F. (2003). Evidence from proteomics that some of the enzymes of actinorhodin biosynthesis have more than one form and may occupy distinctive cellular locations. *Journal of Industrial Microbiology and Biotechnology,* **30**: 523-9.

12. Huang, J., Lih, C.J., Pan, K.H. and Cohen, S.N. (2001). Global analysis of growth phase responsive gene expression and regulation of antibiotic biosynthetic pathways in *Streptomyces coelicolor* using DNA microarrays. *Genes & Development*, **15**: 3183-92.

13. Petrickova, K. and Petricek, M. (2003). Eukaryotic-type protein kinases in *Streptomyces coelicolor* variations on a common theme. *Microbiology,* **149**: 1609-1621.

14. Seifert, H.S., Ajioka, R.S., Paruchuri, D., Heffron, F. and So, M. (1990). Shuttle mutagenesis of *Neisseria gonorrhoeae*: pilin null mutations lower DNA transformation competence. *Journal of Bacteriology,* **172**: 40-46.

Australian Microbial Resources Research Network:

Underpinning Biodiversity, Biodiscovery and Biotechnology Research

L. I. Sly

Australian Collection of Microorganisms, Centre for Bacterial Diversity and Identification, Department of Microbiology and Parasitology, School of Molecular and Microbial Sciences, The University of Queensland, Brisbane

1. Introduction

The vision of this initiative is to develop a world class research network to discover and exploit Australian microbial resources and to make these resources and associated information available for applications in science, research, industry and education. The proposed Australian Microbial Resources Research Network (AMRRN) will provide integrated internet access to Australian collections of microorganisms and electronic access to bioinformation databases to meet national research needs for microbiological resources and to support the competitive development of the life sciences and biotechnology industries in Australia. The network will promote collaborative interactions and accelerate the discovery of Australian microorganisms and microbial genomic information for innovative biotechnology and create new opportunities for bioindustries. The Network will link researchers and foster the discovery and exploitation of Australian microbial resources and make these resources and associated information available for applications in research, industry and education.

2. Microbial diversity

The value of microorganisms as sources of genetic information for molecular biology and biotechnology is well recognized [3,10,12,18], but it was the signing of the Convention on Biological Diversity in 1992 [20] that focussed attention on this area. In the same period the development of molecular methods for detecting microorganisms in the environment revealed the poor state of knowledge of both cultured and non-cultured microbial diversity [e.g. 1,6,9,17,22]. Since then an extensive research effort and substantial research funding has been directed to the areas of microbial diversity, microbial ecology, and biotechnology in Japan, the European Community, and the USA [e.g. 4,18]. North-South attention has also been directed towards collaborations with developing countries in the tropical 'megadiversity' regions and many developing countries have become appreciative of the need to explore, protect, and exploit their own microbial resources. Australia is in a unique position to take advantage of the wide range of ecological habitats of microbial diversity within its own boundaries and in the Asia-Pacific region through collaboration [5]. Access to microbial resources and sovereign rights with respect to microorganisms in environmental samples and cultures in collections have been the focus of several international meetings [7,8,13], and it would be prudent for Australia to develop a national policy in this area as soon as possible.

3. Culture collections in Australia

The status of culture collections in Australia was reviewed in 1998 [15] and it was concluded that there was an urgent need for a network of adequately funded culture collections to conserve and supply Australian microbial resources for use in science, industry and education.

It is essential that microbial cultures are considered as a global resource for the orderly progress of science and technology. However, such a strategy necessitates that each country meets its obligations wherever possible. There are strategic advantages for the *ex-situ* conservation of microorganisms within the country of origin although this may not be practical for some types of microorganisms that require specialist facilities. Many countries have developed national collections to meet their scientific and industrial needs. Australia has depended on institutional collections to meet the needs of their host institutions with little coordination and, except in the plant pathology collections and herbaria, has not recognized the importance of the national perspective. The economic impact on laboratories using reference and type cultures of microorganisms for research and teaching has not been appreciated. In a submission to the National Research Infrastructure Taskforce in 2003 [16] Sly argued for the establishment of Biological Resource Centres incorporating collections of microorganisms to underpin the development of the life sciences, biotechnology and education in Australia. Much of the testing of the microbiological quality of food, water, and pharmaceuticals is carried out in small laboratories that do not have the need, resources or skills to maintain extensive collections of microorganisms and this is best done in specialist Biological Resource Centres. Australian culture collections currently do not have the minimum critical number of staff to undertake the range of conservation, quality assurance, identification, taxonomy, and supply functions which users require [15], but a number of institutional collections could form the core of a national network if adequately funded.

Worldwide, there are almost 500 culture collections of microorganisms registered with the WFCC World Data Centre on Microorganisms (WDCM) [14, 19]. These collections hold nearly one million cultures of microorganisms and cultured cell lines, with by far the majority being held in Europe and North America. In 1998 there were approximately fifty collections in Australia maintaining 65,000 cultures. This number has declined to 35 in 2003.

Culture collections in Australia primarily have institutional roles and the host institutions are usually universities, CSIRO, and government laboratories, together with a few industries. The majority of cultures are bacteria and fungi with minor holdings of protozoa, algae, viruses, plasmids and vectors, and animal cell lines. The major scientific interests of Australian culture collections and their host institutions are listed as general, medical, and agricultural microbiology, followed by veterinary microbiology, microbial taxonomy, plant pathology, applied microbiology, and industrial microbiology. There are also a number of specialist collections engaged in insect microbiology, forest microbiology, food science, and ecology, as well as plant breeding, biodeterioration, mariculture, marine biology, and Antarctic microbiology. Access to information on cultures held in Australian culture collections is extremely limited. Few collections have the resources to publish catalogues and those, which exist, are often out of date. A few collections have developed web accessible catalogues, which is the way of the future. The species directory of Australian culture collections at the WFCC World Data Centre on Microorganisms (WDCM) in Japan provides the most up to date species information but access to strain data of cultures in all the collections is not currently available until collections have the resources to develop web-based catalogues.

4. Australian Microbial Resources Research Network

The development of an Australian Microbial Resources Research Network is seen as a way forward to achieve improved access to information on the location and characteristics of cultures in Australian culture collections and to make these cultures available for research and other applications.

Microbial cultures and associated information databases on phenotypic, genotypic, and ecological characters underpin most areas of research in microbiology, molecular biology and biotechnology and enable progress in these and related disciplines. The key role of culture collections and biological resources has been recognised in recommendations to the Australian government including *The National Strategy on the Conservation of Australia's Biological Diversity, 1992* [21]; *Bioprospecting: Discoveries changing the future* (House of Representatives Standing Committee on Primary Industries and Regional Services), 2001; *Biological Resource Centres: Underpinning the Future of Life Sciences and Biotechnology* (OECD Directorate for Science, Technology and Industry, Committee for Scientific and Technological Policy), 2001 [11]. Implementation of these recommendations is essential for the progress of science and biotechnology in Australia.

The Australian Microbial Resources Research Network [2] has been proposed to develop coordinated national access to comprehensive microbiological and information resources amongst the research community to support the competitive development of the life sciences and biotechnology in Australia for research, industry, and education. The focus of the Network will be on microorganisms, including the Bacteria, Archaea, Micro-algae, Fungi and Yeasts, Protozoa, and Viruses, but may also include collections of animal and plant cell cultures, genetic vectors, and DNA resources, which are maintained using similar methodologies.

The objectives of the Network will be achieved by the development of:

- An internet website to link research groups studying or using Australian microbial resources.
- An Australian Microbial Resources Study (AMRS) to foster collaboration and accelerate the discovery and exploitation of unique Australian microbial diversity.
- An integrated national network of Australian Collections of Microorganisms (ACM).
- An electronic Australian Microbial Resources Information Network (AMRIN) for access to information on microbial biodiversity resources including information on cultures and their characteristics, taxonomic expertise, identification services, ecological data, research facilities, methodologies, relevant publications, and with links to international resources.

- Interactions and collaborations between research scientists and the biotechnology industry for innovative biotechnology.
- Postgraduate and postdoctoral training and mentoring of scientists to meet the national strategic needs for skilled taxonomists, ecologists and microbial biotechnologists.

Network activities will focus on developing strategies to improve information accessibility and exchange between Network participants in order to:

1. Bring improved recognition and a national focus to the study and use of Australian Microbial Resources;
2. Encourage inter-disciplinary research with respect to microbial taxa and research expertise;
3. Facilitate collaborative linkages between researchers and innovative bioindustries.

4.1 Network programs

The proposed Australian Microbial Resources Research Network will link and support high quality researchers working in a range of disciplines including Microbial Diversity, Taxonomy, Evolution and Genomics, Ecology, Identification, Culture Collections, Bioinformatics, Biodiscovery and Biotechnology to promote the discovery and utilization of Australia's unique microbial resources. The Network will deliver its objectives through three programs to consolidate the current fragmented approach and provide a national focus. These programs are the Australian Microbial Resources Study of Australian microbial diversity, an Australian culture collection network to conserve and supply valuable cultures for research and use in science, industry and education, and an Australian Microbial Resources Information Network to facilitate access to information on Australian microbial resources and access to cultures and research expertise to stimulate research and biotechnology and boost collaboration. These programs will provide an impetus to Australian research and development by overcoming long-standing deficiencies in infrastructure which impede researchers' access to microbial cultures and information on these cultures. The integration and coordination of the proposed Network is shown in Figure 1.

The proposed Australian Microbial Resources Study (AMRS) Program will develop strategies to accelerate the description and characterisation of Australian microbial diversity and to progressively make this information available to the research

analysis of data by participants. Mechanisms will be developed for participants to maintain up to date information on their research activities on the AMRIN website.

Progress will be accelerated by overcoming the current fragmented approach and developing a national research focus which recognizes the scientific value of endemic microbial resources and supports the role of culture collections and associated information in this research.

This initiative is truly collaborative in that it brings together the research strengths of participants covering the breadth of Australian microbial diversity and scientists responsible for maintaining valuable collections of microbial resources. Current participants [2] come from most states and territories, and from a range of institutions including universities, research institutes, and Government laboratories with complementary research interests and expertise. This indicates the important role of these resources and scientific expertise across a wide range of disciplines. This also demonstrates wide support for the initiative and the commitment and vision of participants that Australia needs to take a national approach in this area, the absence of which impedes progress compared with Europe, Japan, the USA, and a growing number of developing countries which have recognized the value of national infrastructure for microbial resources and microbial bioinformatics. This view is supported by recommendations made in several key review documents as mentioned previously.

The OECD Directorate for Science, Technology and Industry, Committee for Scientific and Technological Policy recognizes [11] that "Biological Resource Centres are an essential part of the infrastructure underpinning life sciences and biotechnology". Such biological resource centres consist of service providers and repositories of the living cells, genomes of organisms, and information relating to heredity and the functions of biological systems. They include collections of culturable organisms (e.g. microorganisms, plant, animal and human cells), replicable parts of these and viable but not yet culturable organisms (e.g. genomes, plasmids, viruses, cDNA banks), as well as databases containing molecular, physiological and structural information relevant to these collections.

The OECD report [11] emphasises that biological resource centres must meet the high standards of quality and expertise demanded by the international community of scientists and industry for the delivery of biological information and materials. They provide access to biological resources on which research and development in the life sciences and the advancement of biotechnology depends in areas such as the quality of the environment, agriculture, and human health, and for the commercial development of biotechnology [11].

Recognition of the importance of microorganisms as a source of new frontier opportunities for bioindustries in Australia and the need to strengthen and support culture collections of microorganisms to provide the necessary resources to underpin the life sciences and the development of strong and competitive biotechnology has been indicated in authoritative Australian government documents. They not only demonstrate the individual importance of the distinctive fields of microbial biodiversity, microbial resources, bioinformatics and biotechnology, but also that these fields are inexorably interwoven and interdependent. Furthermore, they demonstrate the need for a national resource to integrate and link researchers in the fields and provide access to information as outlined in the proposal.

In the *National Strategy on the Conservation of Australia's Biological Diversity* [21] recommendations were made to strengthen *ex-situ* conservation through the provision of adequate resources to relevant institutions and organizations by: "establishing or strengthening networks of culture collections of microbial species, including those of medicinal, agricultural and industrial importance" and to "accelerate research into taxonomy, geographic distribution and evolutionary relationships of Australian terrestrial, marine and other aquatic plants, animals and microorganisms, priority being given to the least known groups, including non-vascular plants, invertebrates and microorganisms [...]"

Further, the report by the House of Representatives Standing Committee on Primary Industries and Regional Services in August 2001 on *Bioprospecting: Discoveries changing the future*, recognized that in order to overcome impediments in establishing Australian bioindustries, there was a need for increased funding for baseline studies of the Australian biota targeting additional funds for collecting activities in bioactive hot spots; funds for a larger volume of taxonomic work than at present and to ensure sufficient young taxonomists are being trained to undertake this work; and to provide more funding to maintain and expand existing collections so that they provide comprehensive coverage of Australian biota, including microorganisms. In addition, it was recognized that funding for digitizing and networking information about all of Australia's biological resources was required to provide the necessary infrastructure and skills to provide efficient access to information about Australia's biota.

Currently in Australia, there are several institutional collections in Universities, CSIRO, government, and research institutions with the potential to form a national infrastructure of Biological Resources Centres in Australia and many of these forms the core of the current application for the Research Network [2]. However, to be effective, access to these existing facilities needs to be integrated and assistance is required to facilitate data capture and data validation. Leadership is required to develop a national policy and infrastructure support to bring recognition and coordination of these facilities under one umbrella, and the proposed Research Network aims to achieve this.

Developing this infrastructure will enhance current and future research progress in many areas of the life sciences, biotechnology, and education in a way that has been achieved in Europe, Japan, and the USA, which have developed either centralized, or decentralized but coordinated, facilities to support research and innovative bioindustries.

Establishment of an Australian Microbial Resources Research Network will link high quality research and researchers working in a range of disciplines including Microbial Diversity, Taxonomy, Evolution and Genomics, Ecology, Identification, Culture Collections, and Bioinformatics, as well as in Biodiscovery and Biotechnology.

This initiative provides an opportunity to rectify the current fragmented approach and provides a national focus but will need infrastructure funding if it is to achieve its full potential and reach the standard of similar facilities and networks in other countries. If adequately funded it will have a significant impact and provide an impetus to Australian research and development by overcoming current deficiencies in infrastructure and inefficiencies due to major difficulties experienced by researchers to gain access to microbial cultures and information on these cultures.

The proposed Network will thus be an enabling initiative which brings together existing resources, while providing a framework for future development and funding initiatives. It will provide a means to encourage researchers to accession valuable cultures in collections for future research and exploitation, as well as making available reference and control cultures used in standard methods of analysis. Thus the Network will establish essential basic collaborative infrastructure to meet national strategic needs in the life sciences, biotechnology, industry, and educational sectors. It will commence a process to overcome a long recognized deficiency in Australian scientific infrastructure not addressed by previous programs and will foster the sharing of information and resources for innovative research.

Frontier opportunities and the development of a strong and competitive biotechnology industry in Australia will depend on achieving the objectives of this Network to accelerate the discovery of Australian microbial diversity, strengthen and support culture collections of microorganisms, and provide essential information on these resources to researchers and industry.

The Network will provide an excellent opportunity to encourage young researchers. As taxonomic expertise has been run down across the biological disciplines, the Network will become a focus for taxonomic research and provide opportunities for training and collaborative research.

The World Federation for Culture Collections (WFCC) and the OECD Directorate for Science, Technology and Industry, Committee for Scientific and Technological Policy [11] recognize that biological resources in culture collections are a world resource, which needs to be accessible across national boundaries for the orderly progress of science and biotechnology. While Australian scientists have access to international resources, Australia cannot depend on other countries to conserve and maintain its own microbial resources. Issues such as geographic isolation, quarantine regulations, shipping delays, rising international shipping costs, and timely access to biological resources, information and expertise are a national responsibility that can be best resolved by the provision of a high quality national resource.

Many Australian culture collections are already individually Affiliate Members of the WFCCs and provide data to the World Data Centre for Microorganisms (WDCM) in Japan. The Network will increase collaboration with the WDCM by developing electronic links for updating information on Australian microbial resources as novel microorganisms with specific characteristics are brought into culture for the first time. Development of the Network will also increase collaboration with the Global Biodiversity Information Facility (GBIF) to showcase Australian microoorganisms to the global community.

The opportunities facilitated by this initiative will affect the development of research in a wide range of disciplines by providing essential biological resources and services such as:

1. Microbial and cell cultures for applications in industry; biotechnology; human, animal and plant health; research; molecular biology; education; and quality assurance testing.

2. Molecular vectors, genomic DNA, DNA clone libraries, and genetic strains for application in molecular biology research and biotechnology.
3. Internet accessible information on Australian microbial diversity, cultures, and genetic resources.
4. *Ex-situ* conservation of Australian microbial diversity for current and future applications.
5. Expertise and services for isolation of cultures for specific applications; screening of strains for specific traits; identification of cultures for research, industry, testing laboratories, biosecurity; and taxonomic characterization and nomenclature for patents and other purposes.
6. Sources of microorganisms for discovery of new natural product opportunities for bioindustries including novel enzymes, biopolymers, pharmaceuticals, and genes.
7. Sources of microorganisms for use in bioremediation, industrial processes and biotransformations, pollution control, microbial inocula for soil and plant health.

4.2 National benefits and communication

The proposed Australian Microbial Resources Research Network will for the first time in Australian research bring together high quality researchers working in a range of relevant disciplines to support the discovery and exploitation of Australia's unique microbial resources. It will provide an impetus to Australian research and development by initiating a process to overcome current deficiencies that cause major difficulties for researchers to gain access to microbial cultures and information on these cultures in Australia. The Network will thus bring together existing resources, while providing a framework for future development and funding initiatives. It will provide a means to encourage researchers to accession valuable cultures in collections for future research and exploitation, as well as making available control cultures used in standard methods of analysis. Thus, the Network will establish an essential resource to meet national research needs in the life sciences, biotechnology, industry, and educational sectors. It will commence a process to overcome a long recognised deficiency in Australian scientific infrastructure not addressed by previous programs. Without this Research Network culture collection resources will continue to fall behind facilities in Europe, Japan and the USA and seriously affect the competitiveness of Australian research in many disciplines that use or depend on access to microbial resources and taxonomic expertise.

The proposed Australian Microbial Resources Network will make a major contribution to many areas of research. The Network will provide excellent opportunities to make a major contribution to achieving National Research Priorities including *An Environmentally Sustainable Australia* (understanding environmental systems, transforming existing industries, sustainable use of biodiversity), *Promoting and Maintaining Good Health* (diagnosis and prevention of microbial diseases), *Frontier Technologies* (biotechnology, bioremediation, material sciences, nanotechnology, biosensor technology, biomaterials, biopolymers), and *Safeguarding Australia* (critical infrastructure, protecting Australia from invasive diseases and pests, biosecurity).

The Network will develop several strategies to stimulate communication amongst research participants and to facilitate data capture and access to the information. The mechanisms to capture benefits from the Australian Microbial Resources Research Network will include:

- **Communication and Information Exchange**
  - Development of a Network Website to provide access to and dissemination of information for participants and the broad community
  - Establishment of an e-Newsletter for announcements and information exchange
  - Establishment of Advisory Committees to investigate and advise on issues relevant to the operation of the Network and delivery of its objectives
  - Establishment of an annual meeting on Australian Microbial Resources and Biodiscovery for participants and the broader community
  - Holding training workshops to transfer expertise
  - Assist participants to attend relevant conferences and promote Australian microbial resources
- **Facilities and Infrastructure**
  - Providing access to resources and facilities in Australian culture collections by the development of databases and software for data management
  - Providing opportunities for applications for infrastructure funding to benefit Network participants
  - Providing a means to access research expertise

- **Research Collaboration**
  - Fostering opportunities for multidisciplinary research by taking advantage of complementary research expertise
  - Encouraging links between research groups to facilitate innovative research which depends on research expertise in different fields
  - Establishing links with industry to exploit the application of Australian microbial resources in bioindustries by addressing industrial needs and providing industry with potential opportunities.
- **Training**
  - Introducing a mentoring program for young researchers to ensure career development and continuity of research expertise to meet national needs
  - Providing opportunities for young researchers to engage in joint research applications to develop their track record and competitiveness
  - Provide leadership roles for young researchers in symposia, PhD supervision, Advisory Committees
  - Establishing a visiting scientist program to take advantage of distributed research expertise and encourage collaboration
  - Holding technical training workshops to facilitate technology transfer

4.3 Complementarity of the participants

The research interests of the current participants cover a broad spectrum of microbiological research including the microorganisms belonging to the bacteria, archaea, fungi and yeasts, microalgae, protozoa, and viruses. Further details may be obtained by accessing the website [2]. The participants are widely distributed across Australia in Queensland, New South Wales, Victoria, Tasmania, Western Australia and the Northern Territory and are located in universities, Government Departments, and CSIRO. Fields covered include research in human and animal infectious diseases, plant pathology, food science and safety, environmental science, biopharmaceutical discovery, enzyme discovery, molecular diagnostics, microbial identification, and biotechnology applications. In addition, the participants include scientists responsible for a number of the most significant collections of microorganisms in Australia and have linkages to other significant collections which will be included in the Network, including:

Australian Collection of Microorganisms at the University of Queensland

Australian Collection of Extremophilic Microbes at Griffith University

Australian Collection of Plant Pathogenic Bacteria at NSW Agriculture Institute, Orange

Australian Collection of Antarctic Microorganisms, University of Tasmania

CSIRO Collection of Food Fungi and Yeasts, CSIRO Food Science Australia, North Ryde

CSIRO Collection of Living Microalgae, CSIRO Marine Research, Hobart

Plant Pathology Culture Collection, Plant Pathology Herbarium, Queensland Department of Primary Industries

The current participants should be viewed as the initial core of the Network. Already there is significant interest from additional scientists to join the Network and contribute data once it is established. The potential exists to encourage much more extensive participation and to gradually include all significant culture collections in the Network. It is envisaged that the final configuration will include approximately 10 collections in the integrated Australian Collections of Microorganisms.

Acknowledgements

The contribution, enthusiasm and encouragement of the Network participants is gratefully acknowledged in the preparation of this presentation. The financial support of an Australian Research Council Research Network Seed Funding Grant (Project SR0354702) is gratefully acknowledged for the preparation of the Network application and development of the Network website.

References

1. Amann, R.I., Ludwig, W. and Schleifer, K. H. (1995). Phylogenetic identification and *in situ* detection of individual microbial cells without cultivation. *Microbiological Reviews,* **59**: 143-169.
2. Australian Microbial Resources Research Network Website. http://www.cbit.uq.edu.au/amrrn/index.htm
3. Bull, A. T., Goodfellow, M., and Slater, J. H. (1992). Biodiversity as a source of innovation in biotechnology. *Annual Reviews of Microbiology,* **46**: 219-252.
4. Clutter, M. E. (1995). NSF support of biodiversity. *ASM News,* **61**: 319.

5. Hawksworth, D. L. (1994). International initiatives in microbial diversity. In *The Biodiversity of Microorganisms and the Role of Microbial Resource Centres*, (Kirsop, B. and Hawksworth, D. L., Eds.), pp.65-72. World Federation for Culture Collections.
6. Hugenholtz, P. and Pace, N. R. (1996). Identifying microbial diversity in the natural environment: a molecular phylogenetic approach. *TIBTECH,* **14**: 190-197.
7. Information Document on Access to *ex-situ* Microbial Genetic resources within the framework of the Convention on Biological Diversity. World Federation for Culture Collections. (CBD/UNEP/BIO/COP/3/Infodoc.19)
8. Kirsop, B. and Hawksworth, D. L. (Eds.) (1994*). The Biodiversity of Microorganisms and the Role of Microbial Resource Centres*. World Federation for Culture Collections.
9. Liesack, W. and Stackebrandt, E. (1992). Occurrence of novel groups of the domain *Bacteria* as revealed by analysis of genetic material isolated from an Australian terrestrial environment. *Journal of Bacteriology,* **174**: 5072-5078.
10. Nisbet, L. J. (1992). Useful functions of microbial metabolites. *CIBA-Foundation Symposium,* **171**:215-225.
11. OECD*. Biological Resource Centres: Underpinning the Future of Life Sciences and Biotechnology.* OECD Directorate for Science, Technology and Industry, Committee for Scientific and Technological Policy, 2001.
12. Priorities for Microbial Biodiversity Research. Workshop summary and recommendations. Center for Microbial Ecology, Michigan State University, East Lansing, USA (1995).
13. Sands, P. (1994). Microbial diversity and the 1992 Convention on Biological Diversity. In *The Biodiversity of Microorganisms and the Role of Microbial Resource Centres*, (Kirsop, B. and Hawksworth, D. L., Eds.), pp 7-27. World Federation for Culture Collections.
14. Sly, L. I. (1994). Culture collections world-wide. In *The Biodiversity of Microorganisms and the Role of Microbial Resource Centres*, (Kirsop, B. and Hawksworth, D. L., Eds.), pp.29-35. World Federation for Culture Collections.
15. Sly, L. I. (1998). Australian Microbial Resources: Exploring the need for a priority research program on Australian Microbial Diversity incorporating support for a network of Australian Collections of Microorganisms and genetic resources, and an Australian Microbial Resources Information Network.

    http://www.biosci.uq.edu.au/micro/academic/sly/lsly.pdf
16. Sly, L. I. (2003). Biological Resources Centres: Essential Infrastructure for the Life Sciences, Biotechnology and Education. Submission to the National Research Infrastructure Taskforce, Department of Education, Science and Training.

    http://www.dest.gov.au/highered/ri_taskforce/submissions/pdf/r68.pdf
17. Stackebrandt, E., Liesack, W., and Goebel, B. M. (1993). Bacterial diversity in a soil sample from a subtropical Australian environment as determined by 16S rDNA analysis. *The FASEB Journal,* **7**:232-236.
18. Staley, J. T., Castenholz, R. W., Colwell, R. R., Holt, J. G., Kane, M. D., Pace, N. R., Salyers, A. A., and Tiedje, J. M. (1997). *The Microbial World: Foundation of the biosphere*. American Academy of Microbiology.

19. Sugawara, H., Ma, J., Miyazaki, S., Shimura, J., and Takishima, Y. (1993). *World Directory of Collections of Cultures of Microorganisms*. 4th edition. World Federation for Culture Collections World Data Centre on Microorganisms, Saitama.

20. United Nations Environment Programmme (1992). *Convention on Biological Diversity*. United Nations Environment Programme Environmental Law and Institutions Programme Activity Centre, Nairobi.

21. The *National Strategy for the Conservation of Australia's Biological Diversity*. Department of the Environment, Sport, and Territories, Commonwealth of Australia. (1996).

22. Ward, D. M., Weller, R. and Bateson, M. M. (1990). 16S rRNA sequences reveal numerous uncultured microorganisms in a natural community. *Nature (London),* **345**:63-65.

Search for Bioactive Compound Producing Microorganisms

Fungi as a source of natural products

D. Smith, M.J. Ryan and A. Buddie

CABI Bioscience, Bakeham Lane, Egham, Surrey TW20 9TY

1. Introduction

Microorganisms are regarded to have enormous potential to provide solutions to food, industrial, environment and health problems and culture collections play an important role in their conservation and utilisation [39, 65]. Culture Collections have existed for almost as long as the discipline of microbiology itself, indeed public service culture collections go back as far as Kral over 100 years ago [61]. Microbial culture collections provide biological resources (living organisms) and preservation services to microbiologists working in the fields of education, environment, agriculture, and biotechnology. Biotechnology has harnessed living organisms in bioremediation or detoxification of polluted sites, biodegradation of waste, biocontrol, and production of useful chemicals, such as enzymes, biocides, drugs and dyes and in processes such as biotransformations and as food. Microorganisms will provide even more solutions to our problems in the environment, health, agriculture and the economy at a time when natural resources are being depleted and alternative sources of energy and foods are being sought.

However, biotechnology discovery is not simple. It is estimated that there is 1 chance in 250000 for an unknown chemical reaching the market, if you have active compounds this increases to between 1: 5000 to 1: 10 000 [21, 71, 50]. Not only is the chance of getting a product to market low the time frame can be 10-15 years before rewards might be seen. Chances of active molecule discovery are enhanced when fungal metabolism can be manipulated [16], linking this to knowledge of ecosystems, targeted isolation programmes can give access to the most promising organisms. A collection needs to find ways to benefit from the utilisation of its holdings with the goal to provide long-term sustainability for biosystematics, collection maintenance and fundamental research. There are over 470 collections registered with the World Data Centre for Microorganisms (WDCM) of which over 250 hold fungi (**http://wdcm.nig.ac.jp**). There are over 1 million strains held, some 369,119 are fungi representing 23,807 names, but the latter does not take into account the overlap in anamorph and teleomorph names, synonyms and variation in spelling. The strain data on many of these fungi are available via the electronic catalogues provided directly by the individual collections or via national or regional databases such as the WDCM. However, there is a long way to go before we have access to material representing all 100 000 described species and the estimated 1.4 million yet to be described [32]. Molecular technologies will no doubt speed up the process but there will remain the vast majority that will be difficult or impossible to grow in culture. Therefore targeted isolation programmes are needed to make inroads into this enormous task and gain access to this hidden resource. A lot of effort is being put into ecosystem based approaches where little explored or extreme environments are being studied. The Iwokrama programme is an example of such an approach [34]. Endophytic fungi have been isolated from living symptom-less leaves of 12 tree species. The programme isolated 2492 cultures (64 taxa) from which 332 isolates were selected that produced 2800 extracts. As a result 84 isolates were found to have potent anti-insect activity, 14 showed potent anti-fungal activity and 13 potent anti-bacterial activity. A tremendous hit rate of 1 in 3 although not all will be new active molecules and quite often you will find that activity is due to a combination of molecules in the extract and as a result not easily exploited. Such activities can provide income streams to support collections that are part of the discovery process but it needs partnerships an investment to take the molecule to market and without the right skills and technology active molecule discovery remains subject to serendipity.

Key to the use of microorganisms from culture collections is the retention of their properties as research and development must be based on authentic and well-preserved biological material. It is therefore imperative that there is a quality assurance and regulatory framework for the operations of collections. To date several initiatives have lead to the development of guidelines to ensure best practice. The World Federation for Culture Collection (WFCC) have been helping collections in this respect for over 3 decades and have produced *Guidelines for the Establishment and Operation of Culture Collections* (**http://www.wdcm.nig.ac.jp**). In the 90's the UK National Culture Collection (UKNCC) initiative drew together a quality management system (**http://www.ukncc.co.uk**). European collections collaborated similarly to produce the Common Access to Biotechnological Resources Information (CABRI) Guidelines (**http://www.cabri.org**). As a result, strains of organisms are supplied from member collections with traceability, conforming to national and international regulatory requirements, and that are preserved in such a way as to retain their full potential.

The CABI Genetic Resource Collection is a public service fungus culture collection that must take advantage of the potential income streams based upon the utilisation of its holdings. A selected history of its content provides an example of the work that collections do and their important role in the provision of well-maintained authentic cultures, the tools for biotechnology. The CABI collection, originally of the Imperial Mycological Institute (IMI) latterly the International Mycological Institute originated as a dispersed set of example strains about 80 years ago. The collection was created to support agricultural plant pathology in the Commonwealth. This CABI role was organised and strengthened by Major Dade, the first collection curator, and in 1947 CABI took on the additional role of managing the UK National Fungus Culture Collection [1]. The collection is backed up by an extensive dried reference material collection, 'herbarium' of some 380000 specimens that represent over 32000 species. Of these only a small percentage could be grown in culture and 28000 strains representing 6000 species are currently held in the living collection. The most significant accession to the living collection with respect to natural products is IMI 24317, *Penicillium notatum*. Sir Alexander Fleming deposited this as an example of his penicillin producer in April 1945 in the then National Collection of Type Cultures as NCTC 6978. It was taken over in 1948 along with all other non-human pathogenic fungi from the NCTC by CABI and has been kept alive ever since. There is one remaining ampoule that was freeze-dried in October 1944 by Alexander in the USA. The history of strain preservation and handling in the CABI/IMI collection was recorded meticulously. Fleming's strain has been transferred from agar to agar 20 times then placed under mineral oil in 1956 and transferred another 24 times before being freeze-dried in 1976. Since that date this batch of ampoules has been the source of material for subsequent restocking and preservation. The use of freeze-drying and cryopreservation techniques and the establishment of 'seed stock' or 'master stock' limit the change that could occur at each transfer and from long-term adaptation to laboratory conditions. CABI introduced freeze-drying in 1965 and cryopreservation in 1968 to reduce change in preserved materials. CABI uses these techniques as its main methods for storage of fungi and each strain in the collection is preserved by at least one of these methods and one other method. As a disaster measure, the collection is replicated in a building at the other side of the site with strain data stored off site.

Besides Fleming's strain there are several organisms of historical significance including an early example of the fungus that caused the potato famine *Phytophthora infestans*. Other important strains in the collection are the aviation fuel strains, over 40 examples of *Hormoconis resinae* are held, the first one deposited, IMI 88968, was freeze dried in 1968 which coincides with the first occurrences of the fungi as an aircraft fuel problem. The continued presence of this fungus is reflected over subsequent years as strains continued to be isolated and deposited. The continued availability and variety of strains available for test here has led to the development of an early warning test kit (http://www.conidia.com). If gone unnoticed the fungus can bring aeroplanes out of the sky. The Dutch Elm disease fungus is also held in the CABI collection, formerly *Ceratocystis ulmi* now known as *Ophiostoma ulmi.* IMI 185896 is one of 25 strains of this disease causing organism deposited enabling critical comparisons to be made.

The techniques used to preserve the strains are of critical importance and CABI continues to be at the forefront of preservation protocol development. It is essential that the full potential of the strains are retained. For example, the collection holds 130 strains that produce 52 different enzymes including, amylase, cellulase, lipase and proteolytic enzymes. If the organisms were allowed to continue to grow for extended periods without the need for the metabolic pathway requiring a series of enzymes a property could be lost.

The collection holds strains of relevance to most sectors. Plant pathogens and biocontrol agents relevant to agriculture, drug producing strains and opportunistic pathogens of man are important as reference strains for research in medicine. The collection holds 750 strains that produce 442 metabolites that are important to the pharmaceutical sector. CABI also holds many reference strains used in fungal resistance testing to ensure products conform to national and international standards. To date, over 2000 strains have been characterised beyond the information they were deposited with and details of enzyme, metabolite and many other properties have been discovered and recorded [66]. There are strains of value and utility to most industrial, education and research sectors but there is also some way to go. The rapidly advancing methodologies, which incorporate robotics, make the task quicker and easier with metabolomic, proteomic and genomic technologies assisting in natural product discovery. However, these require huge investment.

1.1 The value of microbial genetic resources

Huge investment has been made in the collection, long-term storage, characterisation and maintenance of microorganisms and there are several ways that a financial value can be estimated for culture collection holdings. However, it is not simple, as there are no accepted methods for quantifying the scientific, cultural, educational and other values of a collection. As an example using different criteria an estimated value of the CABI collection holdings has been made. Each culture has a supply value, fungi and bacteria are supplied at £60 per freeze-dried ampoule or £65 for a culture on agar. CABI does not sell the organism *per se* but makes a charge to cover costs of preservation, storage and supply. The cryopreserved collection comprises some 90,000 ampoules that would retail at £5.85 million. There are over 300,000 ampoules containing freeze-dried cultures which would retail at £18 million and over 50,000 universal bottle cultures in mineral oil selling at 3.25 million. This gives the collection stocks a total retail value of £27.1 million. However, it is unlikely that all stocks would be sold, particularly all in one go. It is difficult to ascertain the total number of strains sold world-wide each year but an estimate can be made. There are around 470 collections registered with the World Data Centre for Microorganisms (WDCM). A conservative estimate on sales per collection would be 1 to 4 thousand although there are some large public service collections that far exceed this figure and there are many that supply very few. On this basis it can be estimated that some 0.5 million strains of microorganisms are supplied each year. Selling the entire CABI collection holdings would represent approximately 50% of the world market whilst currently it gets 0.26%.

There are several other mechanisms that could be used for the valuation of the collection including investment costs made up of cost of acquisition, curation and cost of accommodation and alternatively the value may be based on the potential for high value marketable products.

The cost of acquisition is high. The CABI collection was built over a period of 80 years through a combination of CABI scientists going out into the field to collect deposits directly to the collection, strains acquired through the identification service or through specific research projects. The CABI collection holds approximately 6000 species numbering over 28000 strains. The cost at today's prices to isolate, purify, identify and preserve would be approximately £270 per strain giving a total cost to reacquire the collection of £7.56 million.

Isolation and purification	£50
Identification	£110
Preservation by 3 techniques	£100
Data storage	£10
Total	£270

This excludes any form of characterisation, which will inflate these costs depending on the technology and the extent to which the number of techniques used.

However, as many microorganisms have been isolated once or only rarely, some of the holdings of the collection may never be isolated again or at best will take several forays to find. Of the 23807 names of fungi listed in the *World Directory of Collections of Microorganisms* (49%) are listed as present in only 1 of the 250 plus fungal collections in the world (WDCM statistics: http://www.wfcc.info). While these numbers are inflated as they represent names listed in the individual collections, i.e. including some synonyms, anamorph names, and spelling variants, the result is indicative. These data cannot be analysed on a strain basis, but the holdings of CABI may be indicative, of the 6000 species of fungi held 56% are represented by only one strain. Again these figures are distorted, as the accession policy of the collection is one that aims to minimise duplication. If a strain is lost, recovery of that strain or even the same species from its natural environment can be difficult or for practical purposes impossible, unless it is a common organism and has a specific host or environment. This is especially so for strains obtained from soils by dilution methods. Even if the same species is re-isolated, the properties of the strain may be different due to the considerable infraspecific variation that exists. This would inevitably make some strains much more valuable than others. The cost of funding the expeditions to the 142 countries to re-isolate the strains and rebuild the 6000 species collection would also be enormous and is difficult to calculate, but would add several £ million to the cost.

The cost of curation of the CABI culture collection, which includes salaries, cabinets, preservation materials and utility costs exceed £175000. Data is available on collection growth since CABI became the National Collection for Fungus Cultures in 1947. Costs for previous years can be based upon its growth assuming that there is a relationship between this and the cost of curation. The growth of the collection over the period 1947 to 1992 is presented in *IMI: retrospect and prospect* [1]. The collection grew to 10000 from 1947 to 1984 at a rate of approximately 263 strains per year and from 1984 to over 28 000 in 2004 at a rate of 824 per year. On this basis a figure for curation at today's prices for the 57 years would be £5.27 million.

The cost of accommodation can be calculated in different ways, for example a building cost or its rental value. The CABI collection covers less than 8% of the Egham site floor area, the value of the space based on its rental value approximates £60,000 per year and over 57 years at today's prices would cost £3.42 million. The investment cost would be the sum of the cost of acquisition, curation and accommodation giving a figure of £18.25 million (£2 million collection trips; £7.56m isolation and identification; £5.27m curation; £3.42m accommodation).

The final estimate might be made on the fact that the CABI collection could contain organisms that will produce high value products such as drugs, biopesticides, enzymes or chemical leads. The hit rate quoted for screened products to reach the market is varied but because of CABI's unique approach it is likely to hold between 27 and 54 high value products that could reach the market place. Even with low level royalties the potential could be enormous, just 1% of £1 billion per annum is £1 million (see below), and therefore the value of the collection on this basis could also be large. It is apparent based on sale or acquisition the cost to access a strain into the CABI collection approaches £1000. It is not possible to cover such costs by culture sales alone and therefore culture collections must get a share of the value of microbial products.

1.2 The value of products arising from microbial genetic resources

The Biopharmaceuticals Report (http://www.researchandmarkets.com/reports/39083) states that there are over 120 products currently being marketed around the world including nine blockbuster drugs. The global market is currently valued at US$41 billion (£24 billion) growing at a rate of 21% over the past five years. It is considered that the total pharmaceutical market could easily reach US$100 billion (£59 billion) by the end of the decade. 42% of sales of the 25 top-selling drugs world-wide are either biologicals, natural products or entities derived from natural products [71]. However, of every 5000 – 10000 products screened only one becomes an approved drug. Biotechnology may speed bioassays and the production of new drugs, explain more accurately how drugs act in the human system and may reduce the vast costs of pharmaceutical research and development. There are huge revenues from fungal derived drugs e.g. cyclosporin, £0.8 billion: clarithomycin, £0.9 billion and amoxicillin £1 billion. In addition to the value of pharmaceuticals the annual market for industrial detergent enzymes is £0.5 billion. Bioremediation of soil in the EC in 2000 was considered to be worth £40 billion.

The Eighth Annual European Life Sciences Report 2001 produced by Ernst & Young demonstrates the importance of the biotechnology industry in economic growth and human welfare. The report covers only the entrepreneurial life science companies that use modern biological techniques to develop products or services to meet the needs of human healthcare or animal health, agricultural productivity, food processing, renewable resources or environmental affairs. Medical device and large pharmaceutical companies are excluded from the figures. In Europe 105 new companies were established in 2000 and a total income of 8.68 billion Euro (US$ 7.64 billion) realised with a growth rate over the year of 38%. Figures for the USA saw 300 new companies established with a growth rate of 10% and total revenue (again without the large pharmaceutical companies) of US$18.9 billion.

Environmental biotechnology products make it possible to clean up hazardous waste more efficiently by harnessing pollution-eating microbes without the use of caustic chemicals. Industrial biotechnology applications have led to cleaner processes that produce less waste and use less energy and water in such industrial sectors as chemicals, pulp and paper, textiles, food, energy, and metals and minerals. For example, most laundry detergents produced in the United States contain biotechnology-engineered enzymes. The importance of microorganisms as cell factories, producers of active biomolecules is well-documented [78]. Fungi are well known for their capacity to produce bioactive compounds [51] and along with the Actinomycetes (particularly, *Streptomyces* spp.) are the most productive sources of useful biomolecules. As it is not uncommon to find that 15-30% of fungi found in the tropics or in unexplored environments are unknown to science, the potential is enormous. But how does a microbial culture collection capitalise on this and ensure its long-term sustainability?

2. Developing knowledge on strain properties

There are many methods available for screening microorganisms for useful physiological or biochemical properties. Techniques have been derived from a wide range of different biological disciplines including biochemistry, bacteriology, mycology, molecular biology and ecology. Many methods were designed to provide diagnostic characters, but may be extended for use in screening for properties of industrial importance. Procedures can determine a single activity (e.g., urease activity or citric acid production) that can be used to investigate multiple activities (e.g., chromatography to detect a range of metabolites). Many tests are commercially available in easy to use formats (e.g., ELISA kits) that allow standardisation between laboratories. Advances in molecular biology have resulted in a further increase in the variety of methods available for investigating the uses and properties of microbes. Collections can gather initial information that adds value to holdings through the use of growth media tests but to discover, purify and characterise an active molecule can require a huge investment.

Determination of the ability of cultures to grow on specific substrates in media that are defined chemically (e.g. Czapek's medium - in which all the components are known chemicals) or media that are "semi-defined" incorporating large macromolecule compounds (e.g. media containing yeast extract which may differ between batches) provides a simple way in which a variety of bacterial and fungal enzymatic activities can be tested. For example, the ability of a species of fungus to grow on media containing sucrose and nitrate salts as sole sources of carbon and nitrogen implies invertase and nitrate reductase activities. Complex substrates in culture media, often in conjunction with indicators, can be used to screen for a wide range of enzymatic activities in a similar fashion. There are many examples of media formulations that can be used as presumptive tests for specific properties [e.g. 8, 20, 37, 59]. These include detecting fatty acid esterase activity using Tween 80 medium [58], lipase activity using tributyrin lipase medium [40], protease activity using casein hydrolysis medium [59] and amylase activity using starch agar [7].

Tests for tolerance, sensitivity and resistance can be simple to perform and can provide useful information. Compounds can be added to nutrient agar to test the tolerance and sensitivity of particular strains. For example, varying the pH of the media helps to identify and characterise acidophilic and/or alkalophilic organisms. Similarly, growing organisms on nutrient agar that contains high salt concentrations can help screen for halotolerant and halophilic isolates. Variation of temperature, light, aeration and humidity allows assessment of the individual environmental tolerances of strains. Strain resistance to a specific chemical (e.g. heavy metal tolerance) can be tested in the same way.

Biodegradative ability and bioremediation potential can also be simple to detect. Growth on a basal medium containing polycaprolactone diol ("plastic" substrate)

[4, 35] can be used to assess the ability of a fungus to degrade plastic materials. Similarly, other synthetic and organic substrates can be incorporated in growth media to allow an assessment of the biodegradative ability of an isolate. Modifications to this method can allow screening of microorganisms for their ability to remove heavy metals and/or radionucleotides from solution - a process known as "biosorption" [25]. The scope for use of biosorption in microbial remediation of contaminated effluents and soil has been demonstrated using fungi and bacteria [24, 42, 75].

Growing a strain on a Petri dish in the presence of another organism has been widely used in microbiology to study the nature of interaction. This can allow assessment of the antagonistic or synergistic activity of particular strains. Antagonistic properties may be due to chemical or vegetative incompatability. Strains exhibiting antagonistic activity or parasitism may have potential for development as biological control agents or they may have interesting active molecules. In mycology, vegetative compatibility group testing of isolates grown on adapted media (i.e. with nitrate) has been used to characterise many fungi including *Fusarium* spp. and *Colletotrichum* spp. [13, 38, 52). Likewise, interaction tests are useful for genetic and molecular investigations such as mating type experiments [15] and parasexuality studies [31].

In vitro pathogenicity tests are used, commonly, in preliminary studies for the development of biological control agents. The target organisms are inoculated onto test media with the potential control strain (or metabolite/protein extract from the control strain) and the extent of pathogenicity and mortality are recorded. Data obtained from such experiments may aid the assessment of the fitness of potential biological control strains before release into the target environment. Additionally, the data are useful in the assessment of the impact of potential biological control strains against indigenous organisms from the target environment. Pathogenicity tests can also be used to monitor the cultural and physiological stability of an organism and to assess whether an isolate is likely to synthesise enzymes or metabolites that may be of wider importance, during pathogenesis.

There are a variety of commercially available kits for testing carbon source utilisation (e.g., Biolog™ plates and API™ tests). Tests involve either a simple colour reaction that indicates carbon source utilisation (caused by the reaction of a dye with free naphthyl compounds), or simply opacity of the growth medium caused by growth of the organism. Biolog tests (Biolog INC USA) consist of a series of 96-well microtitre plates containing 95 available substrates originally designed for specific groups of bacteria but now available for filamentous fungi (without the tetrazolium dye incorporated in the bacterial plates). Incubation of the plates with a live suspension allows utilisation to be assessed by observation of growth in each well.

The enzymatic activities of microbial cultures may be screened directly by testing for a range of intra- and extracellular enzymes. Such methods were used initially in studies of bacteria and yeasts [e.g. 2, 20] but were later adapted for analysis of fungi where extracellular enzymes have been screened in conidial suspensions and spent culture fluid [12, 45]. The methods can also be used with diluted intracellular extracts. Intracellular enzymes are retained inside the cells and are detected following lysis of the cells and testing the crude or purified extracts.

Specific enzyme activity can be demonstrated using a wide variety of chromophore substituted compounds that are available. Perhaps the most widely used, is the large range of *o*- and *p*-nitrophenyl substituted compounds, where cleavage of the substrate liberates free nitrophenol which is a yellow/brown colour. For example, *o*-nitrophenyl β-D-galactopyranoside has been used as a substrate for β-galactosidase [20]. The enzyme activity cleaves the galactopyranoside leaving free nitrophenol, which is yellow. Fluorogenic substrates such as the 4-methylumbelliferyl (4-MU) substituted compounds also have been used on a wide range of microorganisms [e.g. 3, 6, 18, 27, 48, 60]. The basic principle is the same as for the chromogenic substrates except that the free methylumbelliferone is fluorescent, appearing blue/white under long wavelength (i.e. 365nm) ultra violet light. There are many commercially available 4-MU substrates that allow screening of a diverse range of enzymes (e.g., β-glucosidase, trypsin, and arabinofuranosidase: Sigma, Poole, UK). In addition, 4-methylumbelliferyl-substituted compounds have been used in environmental investigations, for example to measure microbial activity in soil [44]. One of the main advantages of using chromophores and/or fluorophores in analyses is that they may provide quantitative data.

There are several test kits available commercially for testing enzymatic activity. One of the best-known is the APIZYM system (Biomeriuex, S.A. France). This has been used in a number of studies with bacteria [e.g. 41], cell lines [e.g. 47], and filamentous fungi [12, 68], and is used to characterise and test for bacteria from crude extracts [36]. A simple colour reaction indicates enzyme activity due to the reaction of a dye with free naphthyl compounds.

The production of organic acids by fungi has been studied for many years. Extracellular citric acid production by fungi such as *Aspergillus* spp. can be screened in 10-14 day old liquid cultures with Altman reagent [26, 67]. Presumptive positive culture fluids should be confirmed by thin layer chromatography in butanol: acetic acid:water [120:30:50 (v/v/v)]. Altman reagent will give a pink colour with citric acid, but may also react with ketoglutarate, oxaloacetate and glutaconic acid. Additional investigations have elicited methodology for screening for malic acid production by bacteria and fungi [17, 14]; and oxalic acid production by fungi [69].

Electrophoresis can be used to characterise intracellular proteins from a wide range of organisms. The technique involves the separation of proteins through a supporting gel (historically starch but now predominantly polyacrylamide). After separation, the proteins can be identified with specific stains. Some enzymes are present in more than one form (isoenzymes or isozymes) and profiles of these can be used to differentiate genera, species and populations of microbes.

There are many molecular methods available for the analysis and characterisation of nucleic acids. The techniques described have diverse applications including gene cloning, screening of gene expression, species definition, study of interactions, phylogeny, study of degradation processes, and detection of metabolite producers, studies of biodiversity and identification of species in crude samples. General molecular procedures such as restriction endonuclease digestion, cloning, hybridisation and sequencing are applicable to microbial DNA. Molecular biology has been revolutionised by the Polymerase Chain Reaction (PCR) which allows the amplification of large amounts of DNA from very little starting material. Further, it allows screening for specific genes but may also be used for "fingerprinting" genotypes. Common techniques that utilise the fingerprinting-type PCR include RAPDs and AFLP (Arbitrary Fragment Length Polymorphism), and SSCP (Single Strand Conformation Polymorphism). Fingerprinting techniques have allowed scientists to separate taxa that are difficult to distinguish using anatomical or morphological criteria. Non-PCR techniques such as electrophoretic karyotyping of chromosome size and number and genomic digestion by restriction enzymes may also be applied. Other techniques that amplify nucleic acids and not considered in this chapter include the Ligase Chain Reaction (LGR), Nucleic Acid Sequence-Based Amplification (NASBA), Q- beta replicase (Qb) amplification and Strand Displacement Amplification (SDA) [22].

There are a number of fingerprinting techniques that can be used to characterise genetic variations within microorganisms. RAPDs (Random Amplified Polymorphic DNA) have been widely used. RAPDs do not require specific nucleotide sequence details for primer design [22], DNA fragments of undefined length can be amplified (Edel, 1998), only nanogram quantities are required [9] and samples can be processed quickly [10]. However, the number and nature of bands produced are often difficult to interpret [10] and results may not be reproducible because of low stringency of the primers, which emanate from the low PCR annealing temperature. Techniques have been developed that employs the advantages of the RAPD method but with significantly improved reproducibility. The most widely used of these methods has been referred to, variously, as variable number tandem repeat (VNTR) PCR, arbitrarily primed (AP) PCR and simple sequence repeat (SSR) PCR. Primers that had previously been used as fingerprinting probes were adapted for this RAPD-like PCR. Simple sequence repeats, which are complementary to the flanking regions or the core sequence of variable number tandem repeats of the microsatellite DNA [74], were used. Banding patterns were produced that were simple and easier to interpret than RAPD patterns, that could distinguish between strains of the same species and were more reproducible than RAPD fingerprints [10]. Other primers that are commonly used to fingerprint microorganisms include primers that correspond to enterobacterial repetitive intergenic consensus (ERIC) sequences and repetitive extragenic palindromic (REP) elements [22]. A further development has been the advent of inter-single-sequence-repeat-anchored (ISSR)-PCR [79]. This method has the benefit of producing large numbers of fragments throughout the entire genome [70].

Amplified Fragment Length Polymorphism-PCR [73] is a highly reproducible PCR fingerprinting method. Genomic DNA is digested with restriction endonuclease to produce restriction fragments, which are ligated to oligonucleotide “adapters”. A PCR reaction is undertaken with primers that include the adapter sequence and between one and five randomly chosen additional nucleotides. This allows selective amplification of the restriction fragments in which the nucleotides flanking the restriction site match the additional nucleotides of the primer [22].

DNA sequencing is a time-consuming, but powerful technique that allows scientists to read the nucleic acid make-up of individual genes. Sequence information is essential for gene cloning and genetic manipulations. Sequencing can be carried out from known or random amplified samples; this technique is known as SCAR (Sequence Characterised Amplified Region) and can be used to detect sequence variation that is useful for separating isolates at the infraspecific level. In addition, SCAR analysis allows the detection of sequence variation among genes from different organisms. Once sequence data have been obtained, sequences can be compared for similarity with the vast libraries of known sequences that are available over the internet (e.g. at the EMBL or GENBANK databases). Sequencing has been used in systematics to define groupings at the most basic level through the sequences of specific gene regions such as the subunits of the ribosomal gene clusters, allowing proper phylogenetic analyses of taxa to be undertaken.

Several techniques may be employed for the analysis of biological components other than nucleic acids of which chromatography is one of the most powerful. Chromatography is a method for separating metabolites and can also be used for the separation of other biomolecules and chemicals and in the characterisation of individual molecules. Techniques commonly used include thin layer (both normal and reversed phase), preparative layer (normal phase), liquid (normal phase), gas and high performance liquid chromatography.

Colours of spots and the relative distance travelled compared to the solvent front (Rf values) of metabolite spots on TLC plates may be used in an attempt to determine the identity of metabolites by comparing these to available information [e.g., from databases, literature or other sources - see 49]. If Rf values and colours are not available, pure samples of particular fungal metabolites can be obtained from chemical supply companies or from individual laboratories where metabolite research is undertaken (e.g., CABI Bioscience for fungal metabolites). If specific chemical characteristics are known the chemical of interest can be extracted from the producers, although this can be time consuming and experimentally demanding. Once characterised, metabolites can be extracted and purified.

High Performance Liquid Chromatography (HPLC) is an automated system often used for analysing complex mixture arising from fungal extracts. Retention indices may be used between laboratories as a reliable guide to the tentative identification of peaks. There are a number of ways to identify unknown peaks, either by comparing with metabolite standards or by using UV/Vis spectroscopy. The use of diode array detectors increases the amount of data obtained even further.

All cellular life forms possess lipid-rich cytoplasmic membranes. The membrane lipids may be used as markers for the identification and classification of microbes owing to the diversity of their precise chemical structure. Amongst the most important microbial lipid components are amphipathic lipids (e.g. phosphoglycerides and glycolipids) that possess both polar and non-polar regions. The non-polar tails consist, generally, of long chain fatty acid esters [30].

Fatty acid analysis may be considered in the following steps: 1. Cells are grown under standard culture conditions; 2. Saponification of the cells liberates fatty acids from the cell surface; 3. Fatty acids are methylated for greater volatility; 4. Analysis of the fatty acids is by gas chromatography; and 5. Comparisons are made between the test fatty acid profile obtained in 1-4 and a database of profiles of known microorganisms. When grown and harvested under such defined conditions, many species produce their own characteristic profile. This can be exploited as a taxonomic tool [72] and commercial databases such as those from MIDI are available to identify unknown isolates. Whilst the technique has primarily been applied to bacteria, refinements have been made to allow its use with yeasts and fungi [43, 46].

The technologies used to characterise microorganisms are evolving at a rapid pace. Concentration on biomics (proteomics, genomics, metabolomics etc.) has generated a number of extemely powerful technologies that potentially permit screening of large numbers, 'high-throughput' of samples. Methods such as microarray a technique of DNA hybridisation for genome typing and Matrix Assisted Laser Desorption/Ionisation Time Of Flight (MALDITOF) are just two examples of a range of techniques that are constantly developing.

If biomic technologies are applied to characterise organisms, the data generated may be recorded and analysed to identify properties. Once identified, individuals exhibiting a unique property or trait could be further evaluated. For example, an organism known to produce an array of metabolites or enzymes with anti-insect activity may have potential for use as a biological control agent. Undertaking such research may be too much for any single culture collection, but would be achievable if partnerships were forged between the providers of organisms the specialists.

3. Ensuring property stability

A collection must ensure that it preserves its strains in a way that retains their full potential. The technique most often recommended is cryopreservation, which aims to suppress metabolic activity, so that cells have a significantly reduced rate of metabolism or become dormant [63]. There are many useful publications in this area and no need to go into detail here as they have been covered elsewhere [66; 65]. Maintenance of viability and genomic integrity is essential. However, this is not always easy as strains do not always respond well to the same protocols and need special treatment. CABI takes four approaches to develop cryopreservation regimes for recalcitrant organisms: Improvements to traditional methodologies, immobilisation cryopreservation, *in situ* cryopreservation and vitrification cryopreservation.

The application of controlled-rate cooling programmes in association with different permutations of cryoprotectants has been the 'classical' approach to cryopreservation for many years. It remains the most adequate mechanism for the cryopreservation for most microorganisms. A cooling rate of -1°C min^{-1} with glycerol cryoprotectant is suitable for most organisms, but for other organisms (i.e. protoctistan fungi) a cooling rate of -10 °C min^{-1} is often used. Common cryoprotectants include: glycerol, trehalose, dimethylsulphoxide (DMSO), polyvinylpyrrolidone (PVP), desferrisoxamine. Optimal regimes often previously determined using light cryo-microscopy [62] and/or the results from viability assessments. At CABI Bioscience this approach has been used to establish cryopreservation regimes for recalcitrant organisms, specifically *Phytophthora* spp. (a plant pathogen), *Serpula lacrymans* (dry rot fungus), *Diplocarpon rosae* (rose black spot), *Cercospora* spp. and *Moniliophthora* spp. using different cooling rates (1, 10, 30°C min^{-1}) and plunge cooling with either glycerol, DMSO, ethandiol, peptone or inositol as cryoprotectants. Success was measured by analysis of physiology gross culture characteristic and microscopical comparison with pre-preserved wild-types.

Variation of traditional methodologies should be the 'first port of call' before potentially more complex methods of cryopreservation are applied. Minor variations in cooling rate and cryoprotectants can result in significant improvements in viability. This was most apparent with the economically important phytopathogen *Phytophthora*, where a change in cooling regime from cooling 10% glycerol at -1°C min^{-1} to 10% DMSO cooling at -10°C min^{-1} result in a significant improvement in viability. DMSO appeared to be better than glycerol in a number of cases, but its toxicity is also much greater than glycerol [55].

Vitrification, the addition of highly concentrated cryoprotectant solutions to cell-systems that when exposed to ultra-low temperatures initiates the formation of an amorphous glass, prevents the development of damaging ice crystals. Vitrification solutions consist of concentrated combinations of osmotically active compounds (e.g. glycerol, trehalose, ethylene glycol, DMSO & sucrose) and may have the potential to be extremely toxic. Vitrification has previously been applied to animal, plant and human cells and has proved successful for some basidiomycete fungi although post preservation physiological/genomic stability tests and toxicological assessments have not been undertaken. The use of vitrification/cryopreservation has potential for use for a wide range of organisms including oomycete fungi, algae and protozoa. The technique has been successful for the fungi of the genera *Flammulina*, *Boletus* and *Mycena* with no obvious adverse effect on culture morphology [55].

The encapsulation of microorganisms in calcium alginate beads or other appropriate matrix and has been documented as being a suitable mechanism for the preservation of filamentous fungi such as *Serpula lacrymans*, [53]. Once immobilised, specimens can be pre-treated (i.e. osmotically manipulated) before being preserved by traditional preservation methodology. Dehydration before cryopreservation can reduce the potential of damaging ice crystal formation. The properties of the alginate matrix can be altered by varying the chemical concentration of the constituents and by adding "fillers" such as bentonite. The method is straightforward and is relatively cost effective; it also has the added advantage of being applicable to other methods of preservation (e.g. water storage, lyophilisation, oil storage) in addition to cryopreservation as well as having potential for direct storage of environmental samples. One drawback of the technique for microorganisms is the maintenance of sterility; numerous handling 'stages' may increase the prospects of contamination. This can be apparent if the organisms are particularly slow growing as they cannot out-compete the contaminants on revival.

The cryopreservation of an organism in association with its host (or host material) is used for organisms that cannot be cultured using traditional *in vitro* techniques or that cannot be cryopreserved using standard cryo-protocols. *In situ* cryopreservation has potential for a wide range of organisms. For example, obligate pathogens, unculturable organisms or organisms involved in symbiotic associations such as mycorrhizal fungi.

Methods of *in-situ* cryopreservation have immense potential for the storage of microorganisms, particularly obligate pathogens such as microcyclic rust fungi [54]. The use of *in-situ* cryopreservation for protoctistan fungi, embedded in plant seeds has recently been tested with excellent results. The use of *in situ* techniques can therefore by utilised using three approaches, *in situ* cryopreservation with host substrate (for parasites and mutualists), 'donor substrates' or natural environmental substrates. However, this in variably means that methods must be specifically developed for individual organisms.

As a mechanism to assess the effectiveness of preservation and as a quality control measure, culture collections need to employ a standard to assess metabolic integrity. Such tests may include:

- Qualitative methods (observations, microscopy)
- Quantitative methods (growth rate, sporulation capacity)
- Biochemical methods (chromatography of metabolites, enzyme assays & protein analysis)
- Molecular methods (PCR, sequencing, functional genomics)

4. Looking for active molecules

Microbiologists have been characterising microorganisms for decades and along the way have made ground-breaking discoveries not least properties that have saved lives [29], as for example, Fleming's penicillin and many others since then. Large companies and small go through the cyclical trend of revisiting microorganisms and other organisms as sources of natural products [77]. There is not one single approach to natural product discovery but there are lessons to be learned from those experiences that are published and placed in the public domain. An example of this is the Iwokrama programme [34]. The Iwokrama Centre in Guyana is an autonomous not-for-profit research and development institution established by Guyana and the Commonwealth to demonstrate how tropical forest biodiversity may be conserved and sustainably utilised for ecological, social and economic benefits. Although the project entailed more than a bioprospecting programme and the creation of a bioinventory, a goal was to put in place a mechanism to isolate, identify, characterise and produce extracts from microorganisms with activities, particularly anti-insect, anti-fungal and anti-bacterial activities [34].

Key to the success was:

- Selection of ecosystem explored – pristine Guyana Shield Frontier Forest
- Choice of organisms to be targeted – endophytic fungi
- Sampling techniques and analysis carried out by specially trained experts
- Selection of charcterisation techniques as deduplicating tools – molecular analysis included ISSR-PCR and RAPD molecular fingerprinting techniques, rDNA ITS sequencing alongside the morphological and cultural characterisation.

The properties of a fungus are reflected by the habitat in which they grow [76] therefore taking isolates from a wide variety of ecosystems will increase the chances of discovering new products. Hyde [33] describes how the likelihood of novel compound discovery from filamentous fungi can be increased. He stresses the role of the mycologist by identifying the less conspicuous fungi thus increasing the breadth of the screening resource. Additionally, the knowledge of the mycologist helps identify habitats for study in order to target rare or unusual fungi. Concepcion *et al.* [19] review some ways in which screening processes are conducted. Essentially the same key elements followed by the Iwokrama programme are emphasised. They review the common methods for collection and isolation and go on to identify the organism groups that are most likely to provide new compounds e.g. endophytes, ascomycetes and their anamorphs and then recommend the search be conducted in unusual environments. The review includes an extensive list of bioassays with references to where the protocols are described. The bioassays include bioindicators looking for cell function or growth inhibition using viruses, bacteria, fungi, human cell tissue culture, DNA interaction, enzyme inhibition, toxicity assessment, protein binding and specific chemical assays. Collections are encouraged to participate in such processes as this may provide income streams that can contribute to their long-term survival.

The chances of finding active molecules vary enormously but ensuring a targeted and structured approach, the use of appropriate technologies and employment of the right skills it can be rewarding. The Iwokrama programme is a case in point where 1 in 3 organisms showed a potential active product [34] but there are other examples. Bills *et al.* [5] describe recent and future discoveries of pharmacologically active metabolites from tropical fungi. They screened 13000 strains of fungi for antifungal, antibacterial, antibiotics, antivirals, insecticides, antihelminthic agents, cancer, diabetes melitus, inflammation and endocrinology by 2 or 4 methods using 2 extracting solvents. On average 13% yielded active compounds. There is a need to co-ordinate efforts and collections should collaborate to improve their chances of being part of the discovery process whilst continuing the functions were established to perform and being compliant with their international obligations.

5. International obligations

Public service collections have a unique role in the supply of organisms for many different uses and, as all microbiologists should be aware, they must comply with conventions, treaties and national law. This includes the Convention on Biological Diversity (CBD) and the national law that has been put in place to implement it. The CBD requires that Prior Informed Consent (PIC) be obtained in the country where organisms are to be collected before access is granted. Terms, on which any benefits will be shared, must be agreed in advance. The benefit sharing may include monetary elements but may also include information, technology transfer and training. If the organism is passed to a third party it must be under terms agreed by the country of origin. This will entail the use of material transfer agreements between supplier and recipient to ensure benefit sharing with, at least, the country of origin. Many culture collections have operated benefit-sharing agreements since they began, giving organisms in exchange for deposits and re-supplying the depositor with the strain if a replacement is required. Access and benefit sharing rules must be followed and signatory countries to the CBD have agreed a voluntary code of practice, the Bonn Guidelines (http://www.biodiv.org). This means that culture collections must supply strains under material transfer agreements to ensure that the recipient complies with the terms and conditions under which the country of origin allows access. Access legislation and the hope for substantial financial returns from isolated strains are restricting the free deposit of microorganisms in public service collections and the legitimate free movement of strains. Depositors, even today 11 years after the implementation of the CBD, rarely have PIC and under such circumstances culture collections should refuse deposit. However, this would go against the principles on which public service collections were established to facilitate access for the benefit of humankind. Culture collections must assume a proactive role in ensuring both they and the country of origin benefit from exploitation of genetic resources they have helped facilitate legitimate access to. An EU DG XII project, Microorganisms Sustainable Use and Access Regulation International Code of Conduct (MOSAICC) will provide an appropriate code of practice. It has entered a second phase of funding by the European Union and is developing mechanisms to allow traceability and enable compliance with the spirit of the CBD and with national and international laws governing the distribution of microorganisms, whilst not restricting scientific goals (http://www.bccm.belspo.be).

The CBD places obligations on the signatory countries to conserve and sustainably utilise their biodiversity. A driving force for Governmental CBD Focal Points to assist *ex situ* collections meets these national obligations.

References

1. Aitchison, E.M. and Hawksworth, D.L. (Eds.) (1993). *IMI: Retrospect and prospect*. UK: CAB International, p.164.
2. Barnett, J.A., Payne, R.W. and Yarrow, D. (Eds.) (1983). *Yeasts: Characteristics and Identification*. Cambridge University Press, Cambridge, UK.
3. Barth, M.G.M. and Bridge, P.D. (1989). 4-methylumbelliferyl substituted compounds as fluorogenic substrates for fungal extracellular enzymes. *Letters in Applied Microbiology,* **9**: 177-179.
4. Benedict, C.V., Cook, W.J., Jarrett, P., Cameron, J.A., Huang, S.J. and Bell, J.P. (1983). Fungal degradation of polycaprolactones. *Journal of Applied Polymer Science,* **28:** 327-334.
5. Bills, G.F., Dombrowski, A., Paláez, F., Polishook, J. and An, Z. (2002). Recent and future discoveries of pharmacologically active metabolites from tropical fungi. In: *Tropical Mycology*, Vol. 2, *Micromycetes* (Watling, R., Frankland, J.C., Ainsworth, A.M, Isaac, S. and Robinson, C.H., Eds.) CAB International, pp.165-194.
6. Bradbury, J.M. (1977). Rapid biochemical tests for characterisation of the *Mycoplasmatales*. *Journal of Clinical Microbiology,* **5**: 531-534.
7. Bridge, P.D. (1985). An evaluation of some physiological and biochemical methods as an aid to the characterisation of species of *Penicillium* subsection *Fasciculata*. *Journal of General Microbiology,* **131**: 1887-1895.
8. Bridge, P., Buddie, A. and Ryan, M.J. (2001). Characterisation and Screening Methods. In: *The UK National Culture Collection Biological Resource: Properties, maintenance and management*. (Smith, D., Ryan, M.J. and Day, J.G., Eds.). UK National Culture Collection, Egham. p.382.
9. Bridge, P.D. and Arora, D.K. (1998). Interpretation of PCR methods for species definition. In *Applications of PCR in Mycology*, (Bridge, P.D., Arora, D.K., Reddy, C.A. and Elander, R.P. Eds.). pp.63-84. CAB International, Wallingford, U.K..
10. Bridge, P.D., Pearce, D.A., Rivera, A., and Rutherford M.A. (1997). VNTR derived oligonucleotides as PCR primers for population studies in filamentous fungi. *Letters in Applied Microbiology,* **24**: 426-430.
11. Bridge, P.D. and Hawksworth, D.L. (1984). The APIZYM testing system as an aid to the rapid identification of *Penicillium* isolates. *Microbiological Sciences,* **1**: 232-234.
12. Bridge, P.D. and Hawksworth, D.L. (1985). Biochemical tests as an aid to the identification of *Monascus* species. *Letters in Applied Microbiology,* **1**: 25-29.
13. Brooker, N.L., Leslie, J.F. and Dickman, M.B. (1991). Nitrate non-utilizing mutants of *Colletotrichum* and their use in studies of vegetative compatibility and genetic relatedness. *Phytopathology,* **81**: 672-677.
14. Campbell, S.M., Todd, J.R. and Anderson, J.G. (1987). Production of *l*-malic acid by *Paecilomyces varioti*. *Biotechnology Letters,* **9**: 393-398.
15. Casselton, L.A. and Olesnicky, N.S. (1998). Molecular Genetics of mating recognition in basidiomycete fungi. *Microbiology and Molecular Biology Reviews,* **62:** 55-70.

16. Chang, P. K., and Todd, R.B. (2004). Metabolic pathway regulation. In: *Handbook of Biotechnology*, Mycology 20, second edition, (Arora, D.K., Ed.). New York: Marcel Dekker. pp.25-37.

17. Chibata, I., Tosa, T. and Takata, I. (1983). Continuous production of *L*-malic acid by immobilized cells. *Trends in Biotechnology,* **1**: 9-11.

18. Claeyssens, M. (1989). Chromophoric and fluorophoric glycosides as tools in biochemical and biotechnological research. *Applied Fluorescence Technology,* **1**: 11-12.

19. Concepcion, G.P., Lazaro, J.E. and Hyde, K.D. (2001). Screening for bioactive novel compounds. In: *Bio-Exploitation of Filamentous Fungi* (Pointing, S.B. and Hyde, K.D., Eds.), Fungal *Diversity Research Series*, **6**: 93-129.

20. Cowan, S.T. (Ed.) (1974). *Cowan & Steel's Manual for the Identification of Medical Bacteria*. 2nd edition. Cambridge University Press, Cambridge, UK.

21. Crocker, G. (Ed.) (2001). *Ernst & Young's Eighth Annual European Life Sciences Report 2001: Integration*. Life sciences Group. London: Ernst & Young International. p.79.

22. Edel, V. (1998). Polymerase chain reaction in mycology: an overview. In: *Applications of PCR in Mycology,* (Bridge P.D., Arora D.K., Reddy C.A. and Elander R.P., Eds.) pp.1-20. CAB International, Wallingford, U.K.

23. Fort, P., Bonhomme, F., Darlu, P., Piechaczyk, M., Jeanteur, P. and Thaler, L. (1984). Clonal divergence of mitochondrial DNA versus populational evolution of nuclear genome. *Evolutionary Theory,* **7**: 81-90.

24. Gadd, G.M. (1993). Interactions of fungi with toxic metals. *New Phytologist,* **124**: 1-35.

25. Gadd, G.M. (1994). Interactions of fungi with toxic metals. In *The Genus Aspergillus* (Powell, K.A., Renwick, A. and Peberdy, J.F., Eds.), pp.361-374. Plenum Press: New York, USA.

26. Gaffney, G.W., Schneier, K., Diferrante, N. and Altman, K.I. (1954). The quantitative determination of hippuric acid. *Journal of Biological Chemistry,* **206**: 695-698.

27. Grange, J.M. and Clark, K. (1977). Use of umbelliferone derivatives in the study of enzymatic activities of mycobacteria. *Journal of Clinical Pathology,* **30**: 151-153.

28. Grünig, C.R., Sieber, T.N. and Holdenrieder, O. (2001). Characterisation of dark septate fungi (DSE) using inter-simple-sequence-repeat-anchored polymerase chain reaction (ISSR-PCR) amplification. *Mycological Research,* **105**: 24-32.

29. Gutiérrez, S., Cardoza, R.E., Casqueiro, J. and Martin, J.F. (2004). Commercial production and biosynthesis of fungal antibiotics: an overview. In: *Handbook of Biotechnology, Mycology 20*, 2nd edition, (D.K. Arora, Ed.). New York: Marcel Dekker, pp.401-411.

30. Hamilton, R.J. and Hamilton, S. (Eds.). (1992). *Lipid Analysis: A Practical Approach.* IRL Press, Oxford, UK.

31. Hastie, A.C. (1981). The genetics of conidial fungi. In *Biology of Conidial Fungi* Vol. 2 (Cole, G.T. and Kendrick, B. Eds.), pp.511-547. Academic Press, New York, USA.

32. Hawksworth, D.L. (2001). The magnitude of fungal diversity: the 1.5 million species estimate revisited. *Mycological Research,* **105**: 1422-1432.

33. Hyde, K.D. (2001). Increasing the likelihood of novel compound discovery from filamentous fungi. In *Bio-Exploitation of Filamentous Fungi* (Pointing, S.B. and Hyde, K.D. Eds.), Fungal Diversity Research Series **6**: 77-91.

34. Kelley, J., Cannon, P.C., Simmonds, M., Pingal, R. and Simmons, C. (2003). The Iwokrama programme: An approach to the sustainable exploitation of genetic resources. In *Biological Resource Centres and the Use of Microbes* (Lima, N. and Smith, D. Eds.). XXII ECCO Meeting Proceedings. Portugal: Micotecca da Universidade do Minho, pp.185-195.

35. Kelley, J. and Yaghmaie, P.A. (1988). Screening of fungal strains employed in the testing of plastics materials. *International Biodeterioration,* **24:** 289-298.

36. Kilian, M. (1978). Rapid identification of*Actinomycetaceae* and related bacteria. *Journal of Clinical Microbiology,* **8**: 127-133.

37. Klement, Z., Rudolph, K. and Sand, D.C. (Eds.) (1990). *Methods in Phytobacteriology.* Akadémiai Kiadó, Budapest, Hungary.

38. Leslie, J.F. (1987). A nitrate non-utlizing mutant of *Gibberella zeae*. *Journal of General Microbiology,* **133**: 1279-1287.

39. Lima, N. and Smith D. (Eds.) (2003). Biological Resource Centres and the Use of Microbes XXII ECCO Meetining Proceedings. Portugal: Micotecca da Universidade do Minho, p.418.

40. Lima, N., Teixeira, J.A. and Mota, M. (1991). Deep agar-diffusion test for preliminary screening of lipolytic activity of fungi. *Journal of Microbiological Methods,* **14**: 193-200.

41. Ling, W.H., Saxelin, M., Hanninen, O. and Salminen, S. (1994). Enzyme profile of *Lactobacillus* strain GG by a rapid APIZYM system: a comparison of intestinal bacterial strains. *Microbial Ecology in Health and Disease,* **7**: 99-104.

42. McEldowney, S. (1990). Microbial biosorption of radionuclides in liquid effluent treatment. *Applied Biochemistry and Biotechnology,* **26**: 159-180.

43. Marumo, K. and Aoki, Y. (1990). Discriminant analysis of cellular fatty acids of *Candida* species, *Torulosis glabrata* and *Cryptococcus neoformans* determined by gas-liquid chromatography. *Journal of Clinical Microbiology,* **28**: 1509-1513.

44. Miller, M., Palojarvi A., Rangger A., Reeslev M. and Kjoller A. (1988). The use of fluorogenic substrates to measure fungal presence and activity in soil. *Applied and Environmental Microbiology,* **64**: 613-617.

45. Mugnai, L., Bridge, P.D. and Evans, H.C. (1989). A chemotaxonomic evaluation of the genus *Beauvaria*. *Mycological Research,* **92**: 199-209.

46. Muller, M.M., Kantola, R. and Kitunen, V. (1994). Combining sterol and fatty acid profiles for the characterization of fungi. *Mycological Research,* **98**: 593-603.

47. Nardon P., Monget D., Didier-Fichet, M.L. and de The, G. (1976). Comparison of zymogram of three lymphoblastoid cell lines with a new microtechnique. *Biomedicine,* **24:** 183-190.

48. O'Brien, M. and Colwell, R.R. (1987). A rapid test for chitinase activity that uses 4-methylumbelliferyl-N-acetyl-b-D-glucosaminide. *Applied and Environmental Microbiology,* **53**: 1718-1720.

49. Paterson, R.R.M., and Bridge, P.D. (Eds.) (1994). *Biochemical Techniques for Filamentous Fungi*. CAB International, Wallingford, Oxon, U.K.

50. PhRMA (1998). *PhRMA Facts*, Washington DC.

51. Pointing, S.B. and Hyde K.D. (Eds.) (2001). *Exploiting Filamentous Fungi*. The University of Hong Kong: Fungal Diversity Press, p.467.

52. Puhalla, J.A. (1985). Classification of strains of *Fusarium oxysporum* on the basis of vegetative compatibility. *Canadian Journal of Botany,* **63**: 179-183.

53. Ryan, M.J. (2001). The use of immobilisation for the preservation of *Serpula lacrymans*. *Mycologist,* **15**: 66-69.

54. Ryan, M.J. and Ellison, C. (2003). Development of a cryopreservation protocol for the Microcyclic Rust-Fungus *Puccinia Spegazzinii*. *CryoLetters,* **24**: 43-48.

55. Ryan, M.J. and Smith, D. (2003). Preserving a microorganism's capacity to produce active biomolecules. In *Biological Resource Centres and the Use of Microbes* (Lima, N. & Smith, D. Eds.). XXII ECCO Meeting Proceedings. Portugal: Micotecca da Universidade do Minho, pp.197-215.

56. Ryan, M.J., Smith, D., Bridge, P.D., and Jeffries, P. (2003). The relationship between fungal preservation method and secondary metabolite production in *Metarhizium anisopliae* and *Fusarium oxysporum. World Journal of Microbiology and Biotechnology,* **19**: 839-844.

57. Ryan, M.J., Jeffries, P., Bridge, P.D., and Smith, D. (2001). Developing cryopreservation protocols to secure fungal gene function. *CryoLetters,* **22:** 115-124.

58. Sierra, G. (1957). A simple method for the detection of lipolytic activity of microorganisms and some observations on the influence of the contact between cells and fatty substrates. *Antonie van Leeuwenhoek,* **23:** 15-22.

59. Skerman, V.B.D. (Ed.) (1969). *Abstracts of Microbiological Methods*. John Wiley Interscience, London, UK.

60. Slifkin, M. and Gil, G.M. (1983). Rapid biochemical tests for the identification of groups A, B, C, F and G streptococci from throat cultures. *Journal of Clinical Microbiology,* **18**: 29-32.

61. Sly, L.I. and Kirsop, B. (1990). 100 years of culture collections. *Proceedings of the Kral Symposium to celebrate the centenary of the first recorded service culture collection.* Osaka, Japan: Institute of Fermentation, Osaka.

62. Smith, D. and Thomas, V.E. (1998). Cryogenic light microscopy and the development of cooling protocols for the cryopreservation of filamentous fungi *World Journal of Microbiology and Biotechnology,* **14**: 49-57.

63. Smith, D. and Onions, A.H.S. (Eds.) (1994). *The preservation and maintenance of living fungi*. 2nd Edition, IMI Technical Handbooks No 2. pp.1-93. Wallingford, U.K.: CAB International. ISBN 0-85198-902-0.

64. Smith, D. and Ryan, M.J. (2001). Culture collections in the twenty-first century. *Biologist,* **48**: 125-128.

65. Smith, D. and Ryan, M.J. (2004). Current status of fungus culture collections and their role in biotechnology. In: *Handbook of Biotechnology, Mycology 20,* 2nd. edition, (Arora, D. K., Ed.). New York: Marcel Dekker, pp.527-538.

66. Smith, D., Ryan M.J., and Day J.G. (2001). *The UKNCC Biological Resource: Properties, Maintenance and Management*. Egham, Surrey, UK: UKNCC Secretariat. ISBN 0-9540285-0-3, p.382.

67. Smith, I. (Ed.) (1969). *Chromatographic and Electrophoretic Techniques*, Vol. 1. Pitman Press: Bath, UK.

68. St Leger, R.J., Charnley, A.K. and Cooper, R.M. (1986). Enzymatic characterisation of entomopathogens with the API ZYM system. *Journal of Invertebrate Pathology,* **48**: 375-376.

69. Takao, S. (1965). Organic acid production by basidiomycetes. *Applied Microbiology,* **13**: 732-737.

70. Taylor, J.W., Jacobsen, D.J. and Fisher, M.C. (1999). The evolution of asexual fungi: reproduction, speciation and classification. *Annual Review of Phytopathology,* **37**: 197-246.

71. ten Kate, K. and Laird, S.A. (Eds.) (1999). *The Commercial Use of Biodiversity; Access to genetic resources and benefit sharing*. p.398. UK: Earthscan Publications Ltd.

72. Tornabene, T.G. (1985). Lipid analysis and the relationship to chemotaxonomy. *Methods in Microbiology,* **18**: 209-234.

73. Vos, P., Hogers, R., Bleeker, M., Reijans M., Van de Lee, T., Hornes, M., Frijters, A., Pot, J., Peleman. J., Kuiper, M. and Zabeau, M. (1995). AFLP: a new technique for DNA fingerprinting. *Nucleic Acids Research,* **18**: 7213-7218.

74. Welsh, J. and McClelland, M. (1990). Fingerprinting genomes using PCR with arbitrary primers. *Nucleic Acids Research,* **18**: 7213-7218.

75. White, G.F., Snape, J.R. and Nicklin, S. (1996). Biodegradation of glycerol trinitrate and pentaerythritol tetranitrate by *Agrobacterium radiobacter*. *Applied and Environmental Microbiology,* **62**: 637-642.

76. Wildman, H.G. (1995). Influence of habitat on the physiological and metabolic diversity of fungi. *Canadian Journal of Botany,* **73** (Supplement 1), S907-S916.

77. Wildman, H.G. (2003). The fall and rise of natural product screening for drug discovery. *Fungal Diversity*, **13:** 221-231.

78. Wrigley, S.K., Hayes, M.A., Thomas, R., Chrystal, E.J.T and Nicholson, N. (2000). *Biodiversity: New leads for pharmaceutical and agrochemical industries*. UK: Royal Society for Chemistry. p313.

79. Ziÿtkiewicz, E., Rafalski, A. and Labuda, D. (1994). Genetic fingerprinting by simple sequence repeat (SSR)-anchored polymerase chain reaction amplification. *Genomics,* **20**: 176-173.

Australian microfungi as producers of bioactive compounds for drug discovery

R. A. Barrow

Microbial Natural Product Research Laboratory, Department of Chemistry, Australian National University, Canberra, ACT 0200, Australia

1. Introduction

For a great deal of the population, drug discovery is considered a modern phenomenon having evolved from modern science, whereas in reality the concept of drug discovery goes back many centuries and has its origins in nature. The foundations of science are considered to have been laid during the 'Greek Golden Age' which reached its peak around the 5th century BCE. During this period, a new mode of thought concerning nature was developing and this became the foundation of observational science as we know it today, resulting in a more rational and logical approach to medicine and consequently drug discovery.

Empedocles of Agrigentum (504 – 443 BCE) introduced the concept that the four fundamental elements of air, earth, fire and water were the basis of all things which at the time represented a move away from mythical and superstitious reasoning toward logical thought [17]. Hippocrates (460 – 370 BCE), often called the Father of Medicine because of his systematic and rational approach to medicine, was born into this society where reasoning was centred on the four 'humours' as postulated by Empedocles. His methods of observation, scientific assessment and therapeutic practice were the first of their kind and he produced a range of medical texts, including treatments using many naturally derived drugs, that were unsurpassed in their field for many centuries [17]. Theophrastus (372 – 287 BCE), who upon the death of Aristotle (322 BCE) inherited both his library and most importantly botanic garden, is credited with creating the first written herbal, an extensive and precise description of 455 plants and their medicinal properties. Another significant work, which built upon those of his forebears, was *De materia medica* written by Pedanius Dioscorides who lived in the first century of the current era (CE). The work was a study on the preparation, properties and testing of drugs and included over 1000 drugs of natural origin. Dioscorides *De materia medica* was republished many times up until the 16th century and was the basis for the transmission of the recorded knowledge of medicine and drugs throughout the Dark Ages of Europe [29].

Arguably the final great European figure in the drug discovery field before the Renaissance was Claudius Galenus (131-201 CE), better known as Galen. Galen was the first leading figure in experimental physiology and produced more medical texts than any other ancient medical author. He created 'galenicals', mixtures of herbs that were used as remedies until the advent of drugs that consisted of a single agent. With his death in 201 CE, scientific medicine declined in Europe for more than one thousand years [17]. While the Dark Ages in Europe prevailed, the Arab culture of northern Africa, Moorish Spain and the Middle East established comprehensive libraries and founded hospitals and schools of learning thereby advancing the medical knowledge that was to be adopted in Europe with the Renaissance. It was during this time of Islamic science that the drug dispensing houses, pharmacies as we now know them, first came to exist. One of the great Moorish scholars, Abu Mohammed Abdallah Ibn al-Baitar (1188-1248 CE), built on the knowledge recorded by Dioscorides and Galen in the field of medicinal plants. An extensive traveller, al-Baitar wrote a textbook on pharmaceuticals documenting over 1400 drugs [20].

Extensive medical knowledge and the directed use of drugs was not just the domain of the Europeans and the Arabs. Some of the earliest known writings in China were discovered at ruins excavated near Anyang, Henan Province in 1899 CE. Of the thousands of glyphs carved on tortoise shell and animal bone, characters existed for an extensive range of medical concepts. These artefacts, dating from around 1500 BCE, are evidence for the early medical knowledge of the Chinese [41]. Many Oriental texts of both a medical and pharmaceutical nature have been written, documenting the use of many natural remedies, over the last three millennia. Isekiko listed 2605 titles of medical works produced during the era between the Han dynasty and the Sung dynasty (207 BCE to 960 CE). A *materia medica* completed in 659 CE, the *Hsin-hsiu pen-ts'ao* consisted of 850 drugs of which most were plant-derived [36].

1.1 Bioactive compounds

The term 'bioactive compound' is overused by scientists involved in the discipline of drug discovery and represents a fail safe clause covering all and every possible activity imagineable. What exactly is a 'bioactive compound'? According to Philippe Augustus Theophrastus Bombastus von Hohenheim, also known as Paracelsus and not surprisingly where the word Bombastic originates, all substances are poisons. Therefore by definition they are active in biological systems thus making them 'bioactive'.

> *All substances are poisonous; there is none which is not a poison. The right dose differentiates a poison and a remedy.*
>
> Paracelsus (1493-1541)

This definition would appear to somewhat devalue the significance associated with the discovery of a bioactive compound and indeed the statement of Paracelsus that everything is poisonous is given some validity through examination of the data presented below (Table 1). The substances appearing in the table are commonly encountered and each of them, if ingested within a one to two hour period, is capable of causing the death of a small child.

In reality, when searching for a 'bioactive compound' we are invariably searching for a specific activity, whether it is anti-inflammatory, anticancer, antibiotic or some other activity will depend entirely upon the assay(s) in use. In the title of this paper, I am also guilty of the misuse of the term 'bioactive' as while there is no doubt that Australian microfungi produce many compounds active in a wide variety of biological systems, the focus of our research is directed toward the discovery of new antibiotics.

1.2 Rediscovering the definition of 'antibiotic'

When I ask undergraduates to define an antibiotic I am lucky to receive a single correct response. When I ask graduates the same question, it is usually met with some degree of indignation, yet even when pressed seldom is a correct response received. While language evolves and the argument can be progressed that the definition of *antibiotic* has evolved I am always surprised as to how few people know the definition of a class of compounds that changed the face of medicine and continue to be of such prominence both socially and scientifically.

The term antibiotic was first used in 1889 by the French mycologist Paul Vuillemin to describe the ability of one microorganism to destroy another and it was later extended by the British mycologist H. M. Ward to include the antagonistic effects observed among microorganisms [27]. It was however a pioneer of modern medicine, Selman Waksman, who received the 1952 Nobel prize in Physiology or Medicine for the discovery of streptomycin, who in 1941 defined the term 'antibiotic' as;

> *[C]hemical substances that are produced by microorganisms and that have the capacity, in dilute solution, to selectively inhibit the growth of and even to destroy other microorganisms.*
>
> Waksman (1898-1973)

For other compounds, whether they be synthetic or natural, that have the capacity to suppress the growth of bacterial and fungal organisms a perfectly good and well used term already exists. That term being 'antimicrobial'.

1.3 The case for continued antibiotic discovery

As a child growing up in the late 1960s and early 1970s I was constantly reminded by the Daleks, a now extinct alien civilisation from the planet Skaro, that "*Resistance is useless*" [33]. Unfortunately the bacteria and fungi that we co-inhabit this planet with, in their typically recalcitrant manner, chose to ignore this directive. Mankind has always sought to overcome disease and after the discovery of the penicillins by

Table 1: Bioactive compound and amount required to cause fatality in a 2 y.o. infant [8]

Substance (Chemical composition)	Approximate lethal quantity
Water (H_2O)	4 L
Alcohol (CH_3CH_2OH)	100g (about 250mL of whisky or vodka)
Sugar ($C_{12}H_{22}O_{11}$)	300g
Salt (NaCl)	37g
Caffeine ($C_8H_{10}N_4O_2$)	30g (about 20 cups of coffee, tea or 30 cola drinks)
Baking Powder (50% tartaric acid; 50% $NaHCO_3$)	10 g
Crushed Fruit Seeds (cyanide)	5-25 seeds (apple, peach, apricot, plum)
Headache tablets	aspirin 2g , phenacetin (Tylenol) 10 g
Matches ($KHClO_3$)	50-100 match heads
Cigarettes	2 g tobacco, 20 eaten, not smoked. [2 cigarettes contain enough nicotine to be lethal but absorption is too slow.]

Florey and Chain in 1941 [39], briefly gained the ascendency in the epic battle against pathogenic microorganisms. However, this victory was shortlived, and Florey himself recognised and commented upon bacterial resistance to his newly discovered drugs, as early as 1944. The microorganisms have regrouped and developed the ability to defend themselves against the chemical weapons we are exposing them to and are currently outflanking us. Indeed, a recent report published by an expert advisory committee in Australia looking into antibiotic resistance concluded that;

> *Resistance has emerged for all known antibiotics in use. For most antibiotics, and classes of antibiotics, antibiotic resistant genes have also entered the bacterial population in the domains where antibiotics are used.* [35]
>
> JETACAR, 1999

In reality, it has been years of neglect, naïveté, or perhaps a combination of both that has resulted in the therapeutic arsenal that western medicine has relied upon to treat microbial infections to be rendered impotent toward the increasing resistance of pathogenic microorganisms. It is the high use of antimicrobials in the healthcare environment that has led to the inevitable selection for resistant microbes. Multi-resistant *Staphylococcus aureus* (MRSA) is now endemic in most large Australian public hospitals due to the extensive use of methicillin as the drug of choice for the control of staphylococcal infections [23]. Similarly, the extensive use of gentamicin as the drug of choice for broad spectrum empirical therapy of Gram negative bacterial infections has led to the emergence of gentamicin resistant gram negative bacteria (GRGN) such as *Pseudomonas aeruginosa* and *Stenotrophomonas maltophilia*, the latter for which in some cases no known effective antimicrobials exist. The use of broad spectrum antibiotics, such as the 3rd generation cephalosporins and monobactams has increased the risk of resistance developing in non-target bacterial species and the tendency for clinicians to use these drugs, even after specific antibiotic sensitivities have been determined, will inevitably reduce their effectiveness. Resistance to aminoglycosides is also a growing concern. The drug itself is toxic with blood levels needing to be continually monitored, further complicating therapy and contributing to healthcare costs. This has resulted in the use of 3rd generation cephalosporins such as cefotaxime and ceftazadime as an alternative to aminoglycosides further increasing the risk of antibiotic resistance developing in this important group of therapeutic agents.

Over the past 20 years hospital acquired infections have become increasingly more prevalent and difficult to control because of the progressive emergence of resistant microbes. Using data gathered by the United States based National Nosocomial Infections Surveillance System (NNIS) [5]. Figure 1 graphically demonstrates the change in resistance patterns of selected pathogens over a five-year period in the 1990s and contrasts this with the change that can occur in a single year. With many of the commonly prescribed antibiotics now only of limited use, even infections caused by previously innocuous bacteria are now causing patient mortality [16].

Infection and mortality appear to be vastly underestimated by the public. At the start of this century around 2 million patients acquired an infection in direct association with receiving health care in a hospital in the United States. These infections resulted in over 90,000 deaths and had a direct cost to the US economy of around US$5 billion [4]. In Australia around 7000 people die each year as a direct result of acquiring an infection while in hospital [23]. Given that the ratio of deaths in Australia and the United States caused by nosocomial infections approximates the population ratio between the two countries (»1:14), it is enticing to suggest that an estimate of deaths resulting from nosocomial infections could be estimated for any country based on its population, when comparable standards of healthcare are available. Taking the three highest populated nations in Europe this would translate to around 25,000 deaths in Germany and around 18,000 in each of France and Britain, further emphasising the situation we are in with regard to effective antimicrobial therapies.

At present the greatest concern is the resistance that is being exhibited by some strains of bacteria toward the drug vancomycin. In terms of Gram positive infections, vancomycin is often regarded as the final drug of choice when other target specific drugs such as the penicillins have failed. However, Enterococci with intermediate levels of resistance to vancomycin have been described (VRE) and there is evidence that the vancomycin resistance has been detected in staphylococcal strains (VRSA) [9, 11, 24]. In fact, in the last decade of the last century (1989-1999) the percentage of enterococcal infections that were reported to the National Nosocomial Infections Surveillance System that were caused by VRE increased from 0.3% to 24.7% representing an 82 fold rise [12].

Figure 1: Selected antimicrobial resistant pathogens associated with nosocomial infections.

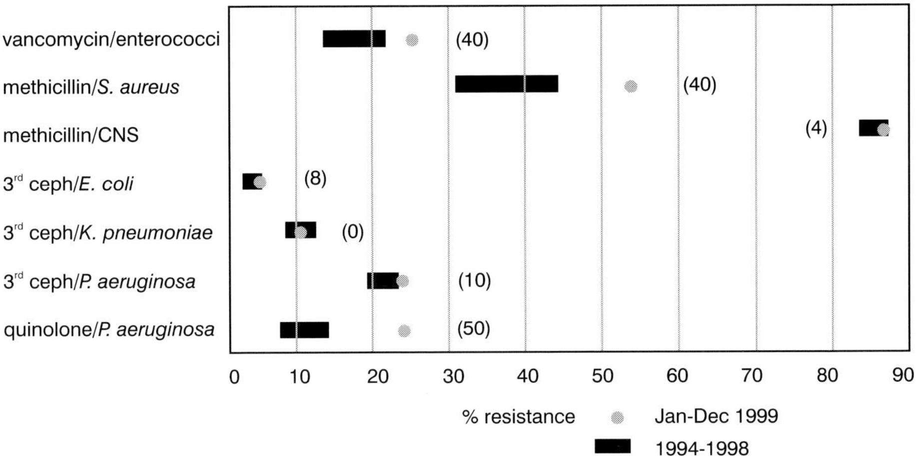

Comparison of resistance rates from January–December 1999 with accumulated 1994-1998 data. Numbers in brackets refer to % increase in resistance for 1999 compared with mean value over previous 5 years. (Source Data: National Nosocomial Infections Surveillance System, Semi-Annual Report 2000).

Key : CNS - coagulase negative staphylococci.

3rd Ceph - third generation cephalosporins (ceftrioxane, cefotaxime or ceftazidime).

Quinolone - refers to either ciprofloxacin or ofloxacin.

Clearly, not only is the health care community concerned by the emergence of VREs, VRSAs and GRGNs but there is also an intense need to define efficacious and less toxic alternatives. While most hospital administrations are acutely aware that the injudicious use of antibiotics can lead to rapid emergence of antibiotic resistance [13], the practices of restricted use of specific agents has only been partially successful with drug resistance in some cases having slowed. However, the loss of efficacy with major drug groups continues with some sense of inevitability and new antibiotics urgently need to be discovered.

2. Fungi in Australia

2.1 The state of taxonomy [34]

It is widely accepted that there are at least 1.5 million species of fungi worldwide [14]. In Australia there is thought to be at least 250000 species of which we currently know only an estimated 14000 (»6%). This estimate of known taxa is based on Pascoe's [25] estimate of 12500 species known in 1990 and extrapolated using the figure of 100 newly described Australian taxa occuring each year [31]. At the current rate of description it will take over 2000 years to entirely catalogue the Australian fungi, however given that there are only around 40 people in Australia with an interest in fungal taxonomy and that most Australian universities no longer employ taxonomic mycologists capable of training the future generations of taxonomists the time estimate is likely to get larger [18].

In the Microbial Natural Product Laboratories we consider that taxonomy is a critical factor in our study, from a basic science perspective in documenting the biodiversity of Australian microfungi and from an applied perspective to assist in our dereplication strategies. Fortunately fungal taxonomy has been revolutionized by molecular technology over the past few years and through collaboration we have started to use molecular techniques to identify fungi. DNA has been extracted from fungal mycelia; PCR amplified using primers to internal transcribed spacer (ITS) regions of rDNA and then separated on electrophoretic gels. The DNA has then been isolated, purified and sequences determined [1]. Comparison with pre-existing fungal sequences in GenBank has revealed sequence homologies for a number of isolated fungi with named species. In addition a number of isolates have had no significant homology with pre-existing fungal sequences and thus may represent new species. PCR amplification of selected segments of ribosomal DNA are particularly useful for fungal taxonomy as they are rapidly evolving and can vary quite significantly between fungal species [40].

2.2 Is there endemism in fungi?

Australia is well recognised for its unique floral and faunal endemism. While the applicability of mycogeography is limited due to sparse knowledge of fungal biodiversity, there is little doubt that Australia possesses many endemic fungi [10]. For example in the genus *Amanita*, which is truly cosmopolitan, 14 out of 16 new additions made to the Australian records since 1980 are species unique to Australia [31], while for the truffle-like fungi the endemism is apparent at the genus level [18].

Arising out of this endemism is a unique opportunity for chemistry as it pertains to drug discovery. Access to novel fungi will give rise to a wealth of new molecular architectures and associated biological activities in the form of secondary metabolites they produce. When one considers the amazing structural diversity and many successful drugs developed from the small sample of the Fungal Kingdom studied so far, none of which are Australian, the potential for Australian derived fungi to produce future drug candidates is truly outstanding.

2.3 Fungi as producers of antibiotics and other bioactive metabolites

The recorded history of fungal medicinal usage dates back thousands of years. At least three millennia ago, the Mayans in Central America were using a fungus they called "cuxum" which grew on roasted corn to treat ulcers and intestinal infections [3]. At around the same time, moulds were used by the Chinese to treat skin and wound infections and by Indian physicians in the treatment of dysentery [37]. As noted previously, as medical knowledge and methods for recording information developed, a number of *materia medica* began to appear in various parts of the world. The Ebers Papyrus, written in Egypt around 1550 BCE, is one such text and contains over 800 prescriptions of predominantly natural origin. It makes frequent mention of yeasts and mouldy bread for treating skin abrasions, infections and eczemas [37].

Given the historical use of fungi and their role as medicines, it is somewhat surprising that until recently fungi were placed in the same kingdom as plants. Today however they are recognised to have representatives spanning three of the six Kingdoms and are found across four Divisions in the Kingdom Eumycota, where they are often referred to as the true fungi, in the Kingdom Protoctista in a single Division and in the Kingdom Chromista, where they are represented across two Divisions. In reality the fungi have more in common with animals than plants and as such the mechanisms they have evolved to defend themselves against threats posed by antagonistic microorganisms frequently have relevance against microbial pathogens of animals, for example those in the suborder Anthropoidea of the class Mammalia, namely humans. Many of our most important medicines have been discovered through programs of research focused toward fungi [6]. Drugs such as the penicillin antibiotics, surely among the best known of the ethically prescribed pharmaceuticals, the antibiotic cephalosporins, the immunosuppressive agent cyclosporin A, the antimycotic agent amphotericin A or the most profitable drugs of recent years, the cholesterol lowering statins, such as simvastatin [19], exemplify the position of central importance that fungi hold today in modern medicine.

Yet, despite the success of exploiting fungi as a source for lead discovery, it is estimated that only around 5% of all fungal species have been described. From the antimicrobial perspective history shows that the best source of drugs has been and continues to be from natural sources with over 60% of currently used antibiotics being derived from natural product leads [7]. However, the past two decades have witnessed a withdrawal in natural products discovery programs in favour of combinatorial chemistry approaches based primarily on the ability of combinatorial chemistry to feed the demand of high throughput screening programs. Today, due in no small part to the extraordinary lack of success delivered by *de novo* combinatorial chemistry which has failed to produce a single drug in the past 23 years coupled with the alarming decrease in new drug applications [30], a renaissance in the proven strategy of natural product chemistry is occurring. Researchers in pharmaceutical companies and academia alike are now working smarter with a focus increasingly on microbial domains. Molecular based taxonomic techniques are being used to assist dereplication efforts, metagenomics is allowing access to the gene products of unculturable organisms, while advances in technology has permitted structures to be solved faster than ever before.

3. The Microbial Natural Product (MINAP) Research Laboratory at ANU

3.1 The MINAP collection

The Microbial Natural Product research laboratories were established in the Department of Chemistry at the Australian National University at the start of this century. During the past four years a library of microfungi has been established with accessions isolated from four ecological niches:

i) Marine fungi from southeastern NSW coastal waters.

ii) Entomogenous fungi from Australian insects.

iii) Endophytic fungi, exclusively from Australian native plant species.

iv) Coprophilous fungi from Australian native animal scat.

The MINAP collection comprises over 1000 fungal cultures and efforts are now underway to document these using molecular techniques. This will not only assist in our own internal dereplication efforts but would substantially enhance the knowledge base of Australian biodiversity as it pertains to microfungi.

3.2 The screening protocol

The primary focus within the laboratories is on the detection of fungal extracts that have the capacity to halt the growth of nosocomial pathogens. The procedure employed is the disc diffusion susceptibility assay. Various modifications to the original method [2] have been described and the procedure actually utilised, for antibacterial assays, is that described by the National Committee for Laboratory Standards (NCCLS) as detailed in their Standard M2-A7 [26]. For antifungal assays, in which *C. albicans* and *A. fumigatus* are screened as the model species, we utilise the NCCLS reference method for broth dilution antifungal susceptibility testing (M27-A2) [28]. The NCCLS method employed in our screening episodes has been developed to standardise antibiotic susceptibility testing and is based on specific protocols and the use of NTCC strains to determine cut off levels for specified antibiotics.

Selection of test microbes (Table 2) has been made on the basis of their clinical importance and propensity to become or be multi-resistant to commonly used antibiotics in the nosocomial setting. NTCC strains are used for control and standardisation purposes, while clinically derived duplicates of these organisms are used for test purposes. The clinical isolates are characterised and have had their antibiotic susceptibility patterns determined prior to use.

3.3 Biocombinatorial approach

The techniques utilised for the culturing of fungi within the laboratories are generally standard and are described in detail in many texts. In addition to standard protocols, the research seeks to develop a biocombinatorial approach invoking both chemical and biological elicitation in efforts directed toward the diversification of metabolite architecture from within a single accession.

It has long been recognised that secondary metabolite architecture can be altered by adjusting the chemical balance of the culture medium. For example by controlling the concentration of ammonium and phosphate ions within a fermentation broth increased yields and structural diversity in metabolite architecture have been

Table 2: Microbial pathogens employed in initial screening program

(-) *Pseudomonas aeruginosa (GS)*	(+) *Staphylococcus aureus (MS)*
(-) *Pseudomonas aeruginosa (GR)*	(+) *Staphylococcus aureus (MR)*
(-) *Pseudomonas cepacia (GS)*	(+) *Staphylococcus epidermidis*
(-) *Pseudomonas cepacia (GR)*	(+) *Enterococcus faecalis (VIS)*
(-) *Enterobacter cloacae*	(+) *Streptococcus pneumoniae (IPRS)*
(-) *Citrobacter freundi*	(+) *Streptococcus pneumoniae (PSS)*
(-) *Klebsiella pneumoniae*	(+) *Staphylococcus saprophyticus*
(-) *Stenotrophomonas maltophilia*	(+) *Mycobacterium smegmatis*
(F) *Aspergillus fumigatus*	(F) *Candida albicans*

GS-gentamicin sensitive; GR-gentamicin resistant; MS-methicillin sensitive; MR-methicillin resistant; VIS-vancomycin intermediate sensitive strain; IPRS-intermediate penicillin resistant strain; PSS-penicillin sensitive strain; (+)- Gram-positive bacterium; (-)- Gram-negative bacterium; (F)- fungus.

observed. The addition of magnesium phosphate, an ammonium ion sequestering agent, has been demonstrated to substantially increase the yield of macrolide antibiotics [15], and we have found that it also impacts on metabolite architecture allowing previously unobserved metabolites to be isolated. Another strategy we adopt involves the addition of magnesium carbonate, well known to deplete phosphate ion concentration, to the culture medium [15, 32]. While high phosphate ion concentration will suppress the production of most metabolites within a culture the sensitivity of production can differ between metabolites as a function of phosphate levels. We routinely explore *chemical elicitation* methods for the diversification of metabolite composition within an active accession and monitor altered metabolite production, which are expressed in yield or architectural changes, by analytical HPLC methods. Recently we have shown that a *Penicillium* sp. (MINAP-9902) can be made to produce sclerotiorin (**1**) and related metabolites (**2**-**4**) or alternatively through the manipulation of media chemistry the production of these metabolites can be effectively suppressed and the production of a previously unobserved metabolite, a tetramic acid we have named ravenic acid (**5**), can be reliably induced (Figure 1) [21, 22].

Figure 1: Metabolites of *Penicillium* sp (MINAP-9902). Sclerotiorin (1), isorotiorin (2), ochrephilone (3), isochromophilone I (4), ravenic acid (5)

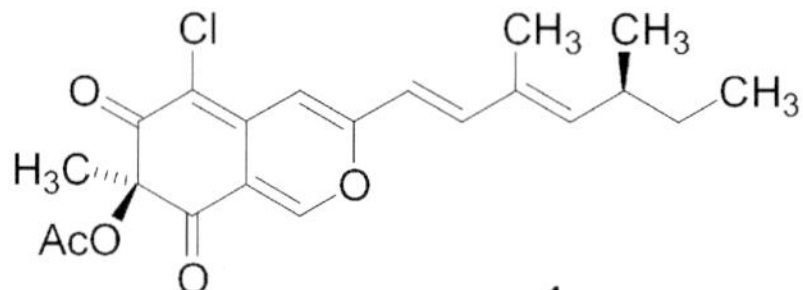

1

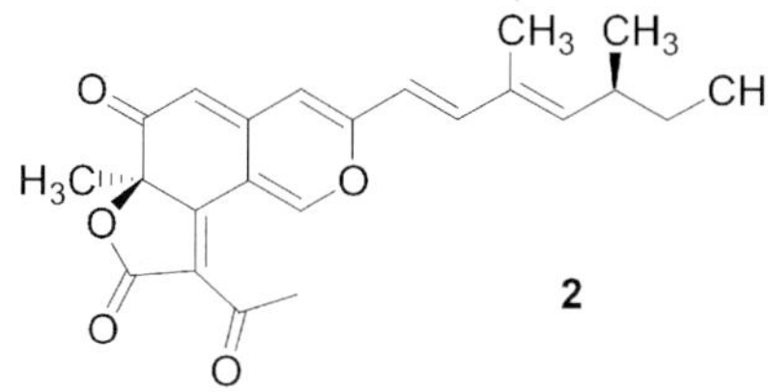

2

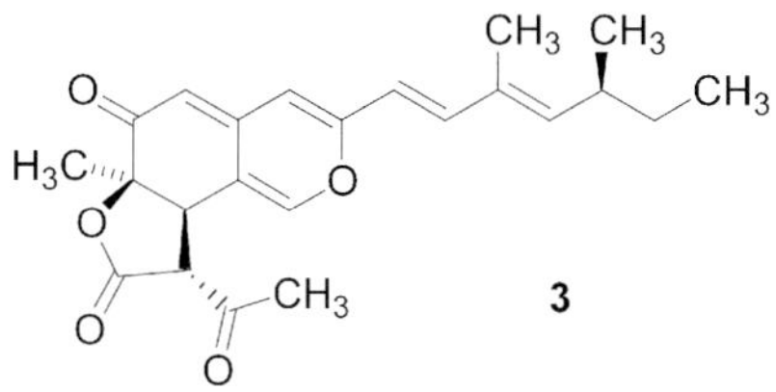

3

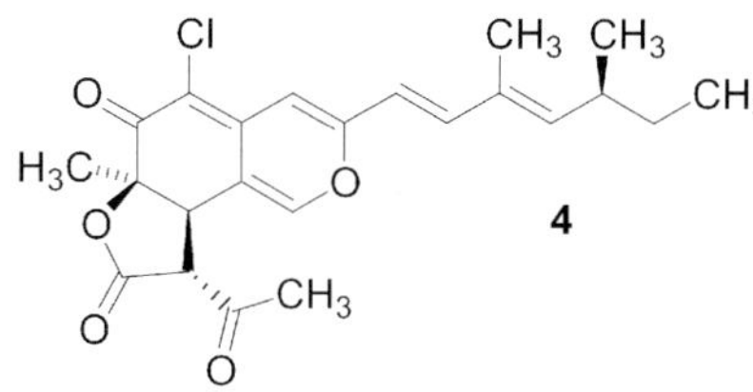

4

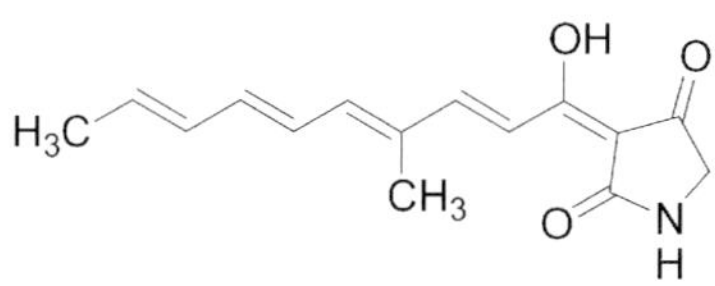

5

In addition to the chemical elicitation techniques employed, biological elicitation is also utilised in our biocombinatorial approach in further efforts to both increase the yield of metabolites and to encourage metabolite diversity.

In order to influence biochemical pathways operating within a cultured microfungus non-viable cells of bacteria (*E. coli* and *S. aureus*) and a fungus (*C. albicans*) are used to supplement the growth medium. In this way the expression of dormant genes responsible for antibiotic production may be elicited [38]. We have recently been successful at substantially increasing the yield of a targeted metabolite, xanthoquinoidin A1 (**6**), required for semisynthetic studies. Yields were increased around 500% by addition of non-viable cells of *S. aureus* to the culture medium (Figure 2). Further to the increased levels of production of xanthoquinoidin A1 we have also observed the production of hitherto unrecognised metabolites from this fermentation. While at this stage these metabolites remain unidentified, it does further validate our biocombinatorial approach to discovery. Whether this occurs through recognition and response to the presence of complex chemicals such as the cell wall material of a potential pathogen or through the presence of a simple pleiotropic effector such as A-factor (2-isocapryloyl-3- *R*-Hydroxymethyl-α-butyrolactone) or a similar acting compound is not known.

To date, due to resource allocation, this biocombinatorial approach has only been implemented on cultures that have shown activity in our antibiotic screening program. In the future it would be useful to reculture accessions that did not demonstrate activity in the initial screening program employing our protocols of chemical and biological elicitation and reassess activity.

Figure 2: Xanthoquinoidin A1 (6) from a *Chaetomium* sp (MINAP-0017).

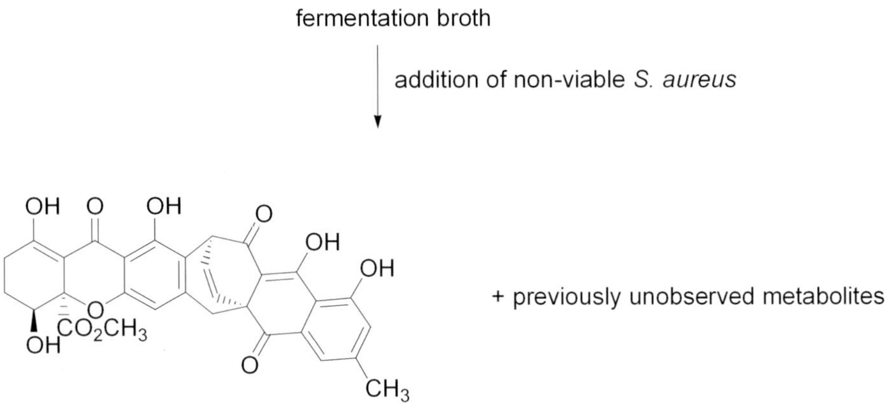

xanthoquinoidin A1 (6)

Acknowledgements

It is with pleasure that I am able to acknowledge the contributions of my collaborators and past and present research students. Special thanks are extended to Dr Mike Kotiw and Sally Kingsland for their contributions toward the creation of this paper.

References

1. (a) Agerer, R., Ammirati, J., Baroni, T. J., Blanz, P., Courtecuisse, R., Desjardin, D. E., Gams, W., Hallenberg, N., Halling, R., Hawksworth, D. L., Horak, E., Korf, R. P., Mueller, G. M., Oberwinkler, F., Rambold, G., Summerbell, R.C., Triebel, D. and Watling, R., (2000). Always deposit vouchers. *Mycological Research,* **104**: 642-644. (b) Selosse, M. A., Weiß, M., Jany, J. L. and Tillier, A., (2002). Communities and populations of sebacinoid basidiomycetes associated with the achlorophyllous orchid *Neottia nidus-avis* (L.) L.C.M. Rich. and neighbouring tree ectomycorrhizae. *Molecular Ecology,* **11**: 1831-1844.

2. Bauer, A.W., Kirby, W.M.M., Sherris, J.C. and Turck, M. (1966). Antibiotic susceptibility testing by a standardized single disk method. *American Journal of Clinical Pathol*ogy, **45**: 493-496.

3. Buss, T.; Hayes, M. A. (2000). In *Biodiversity: New Leads for the Pharmaceutical and Agrochemical Industries* (Wrigley, S. K., Hayes, M. A., Thomas, R., Chrystal, E. J. T. and Nicholson, N., Eds.). Royal Society of Chemistry: Cambridge, UK. Special Publication No. 257.

4. *CDC Morbidity and Mortality Weekly Report*, **49**:8: 149-153. (2000). See also *CDC Morbidity and Mortality Weekly Report*, **49**: 7: 138. (2000).

5. Centers for Disease Control and Prevention (CDC), US Department of Health and Human Services, Public Health Service, National Nosocomial Infections Surveillance System (NNIS), Semi-Annual Report, December 2000.

6. Cordell, G. A., (2000) Biodiversity and Drug Discovery - A Symbiotic Relationship. *Phytochemistry*, **53**: 463-480.

7. Cragg, G. M., Newman, D. J. and Snader, K. M., (1997). Natural products in drug discovery and development. *Journal of Natural Products,* **60**: 52-60.

8. Data adapted from the National Chemistry Week Handbook (1984), *Everything is Toxic: It Depends on the Dose*, by M.G. Hogben, Dept. of Chemistry, Concordia University, Nov. 7. The average weight of the 2 y.o. infant was taken as 10 kg or about 23 lb. http://www.cheminst.ca/ncw/handbook/eng/html/every_toxic_8.htm

9. Emergence of Vancomycin Resistant Enterococci in Australia', Report by NHMRC Working Party on Antibiotics, November, 1996. http://www.health.gov.au/nhmrc/publications/reports/cd12rep.htm

10. For a general discussion of fungal endemism as it pertains to Australia see: http://www.anbg.gov.au/fungi/mycogeography-nearby.html#endemic.

11. For further information the reader is directed to the CDC website: http://www.cdc.gov/ncidod/hip/Aresist/aresist.htm

12. Fridkin S. K. and Gaynes R. P. (1999). Antimicrobial Resistance in Intensive Care Units. *Clinics in Chest Medicine*, **20**:2: 303-316.

13. Geue, A. (2003). Communicable Diseases Intelligence-supplement, Communicable diseases Australia, Department of Health and Aging, May 2003.

14. Hawksworth, D. L. (1997). Fungal and International Biodiversity Initiatives. *Biodiversity and Conservation,* **6**: 661-668.

15. Lancini, G. and Lorenzetti, R. (Eds.) (1994). *Biotechnology of Antibiotics and Other Bioactive Microbial Metabolites*, Plenum Publishing Corp.

16. Levy, S. B. (1998). The Challenge of Antibiotic Resistance. *Scientific American,* **278**(3): 46-53.

17. Mann, R. D. (Ed.) (1984). *Modern Drug Use An Enquiry on Historical Principles*. MTP Press: Lancaster, England.

18. May, T. W. and Pascoe, I. G. (1996). History of Taxonomic Study of Australian Fungi. In *Fungi of Australia,* Vol. 1A Introduction – Classiifcation, ABRS/CSIRO, pp.71-206.

19. *Med Ad News*, Vol. 17, May 1998, p.18.

20. Mez-Mangold, L. (Ed.) (1971). *A History of Drugs*; F. Hoffmann-La Roche: Basle, Switzerland.

21. Michael, A.P., Grace, E.J., Kotiw, M. and Barrow, R.A. (2002). Ravenic acid, a new tetramic acid isolated from a cultured microfungus, *Penicillium* sp. *Journal of Natural Products*, **65**: 1360-1362.

22. Michael, A.P., Grace, E.J., Kotiw, M. and Barrow, R.A. (2003). Isochromophilone IX, a novel GABA-containing metabolite isolated from a cultured fungus, *Penicillium* sp. *Australian Journal of Chemistry,* **56**: 13-15.

23. National Surveillance of Healthcare Associated Infection in Australia', A Report to the Commonwealth Department of Health and Aged Care, April 2001. see http://www.health.gov.au/pubhlth/strateg/jetacar/pdf/scope.pdf

24 a. Padiglione A. A., Grabsch E. A., Olden D., Hellard M., Sinclair M. I., Fairley C. K. and Grayson M. L. (2000). Fecal Colonisation with Vancomycin-Resistant Enterococci in Australia. *Emerging Infectious Diseases*, 534-536. (b) 'Some *Staphylococcus aureus* may be resisting vancomycin', Outreach, July 1997, http://www.psych-health.com/staph.htm; (c) Handwerger, S., Raucher, B., Altarac, D., Monka, J., Marchione, S., Singh, K. V., Murray, B. E., Wolff, J. and Walters, B. (1993). Nosocomial outbreak due to *Enterococcus faecium* highly resistant to vancomycin, penicillin and gentamicin. *Clinical and Infectious Disease,* **16**: 750-755, (d) Rubin, L. G., Tucci, V., Cercenado, E., Eliopoulos, G. and Isenberg, H. D. (1992). Vancomycin-resistant *Enterococcus faecium* in hospitalized children. *Infection Control and Hospital Epidemiology,* **13**: 700-705. (e) Veach, L. A., Pfaller, M. A., Barrett, M., Koontz, F. P. and Wenzel, R. P., (1990). Vancomycin resistance in Staphylococcus haemolyticus causing colonization and bloodstream infection. *Journal of Clinical Microbiology,* **28**: 2064-2068.

25. Pascoe I. G. (1990). History of Systematic Mycology in Australia. In *History of Systematic Botany in Australasia* (P.S. Short, Ed.), pp.259-264.

26. Performance standards for antimicrobial disk susceptibility tests; Approved Standard 7th edition. M2-A7.' National Committee for Clinical Laboratory Standards, (2000). NCCLS, Wayne, PA.

27. Pramer, D. (1988). Some Choice Words from Waksman. The Scientist, **2**[7]: 17 Apr. 04. http://www.the-scientist.com/yr1988/apr/pramer_p17_880404.html

28. Reference method for broth dilution antifungal susceptibility testing of yeasts; approved standard NCCLS document M27-A2.' National Committee for Clinical Laboratory Standards, (2002). NCCLS, Wayne, PA.

29. Riddle, J. M. (Ed.) (1985). *Dioscorides on Pharmacy and Medicine*; University of Texas Press: Austin.
30. Rouhi, A.M. (2003). Rediscovering Natural Products. *Chemical & Engineering News*, **13**: 77-91.
31. Scott, G. A. M., Entwistle, T. J., May, T. W. and Stevens, G. N. *A Conservation Overview of Australian Non-marine Lichens, Bryophytes, Algae and Fungi*, Australian Government, Dept. of Environment and Heritage. (www.deh.gov.au/biodiversity/threatened/action/cryptogams/10.html
32. *Splendid Gifts from Microorganisms – The Achievements of Satoshi Omura and Collaborators.* (The Kitasato Institute, Ed., 1998).
33. The Daleks are a creation of Terry Nation and appeared in the long running BBC-TV series *Dr Who* that ran from 1963-1989.
34. The reader is also directed to http://www.anbg.gov.au/fungi/history-today.html for further information regarding the history of taxonomic mycology in Australia.
35. The Use of Antibiotics in Food Producing Animals: Antibiotic-resistant Bacteria in Animals and Humans' (1999). Joint Expert Technical Advisory Committee on Antibiotic Resistance (JETACAR) Canberra. October 1999. Commonwealth Department of Health and Aged Care. see http://www.health.gov.au/pubs/jetacar.htm
36. Unschuld, P. U. (Ed.) (1986). *Medicine in China A History of Pharmaceutics*. University of California Press, Berkeley.
37. Welch, H. and Martí-Ibáñez, F. (1960). *The Antibiotic Saga*. Medical Encyclopaedia, New York.
38. While elicitation is well documented in higher plants (e.g. phytoalexin production) the method in microorganisms is relatively unexplored. For paper in elicitation as applied to cyanobacterial cultures see Patterson, G.M.L. and Bolis, C.M., (1997). Fungal cell-wall polysaccharides elicit an anti-fungal secondary metabolite (phytoalexin) in the cyanobacterium *Scytonema ocellatum*. *Journal of Phycology*, **33**:1: 271-276.
39. While Fleming shared the Nobel Prize with Florey and Chain in 1945 for the discovery of penicillin, his work was very much earlier and consisted of recognising the potential of the fungus *Penicillium notatum* to produce compounds active against pathogenic microorganisms. It was the pioneering work done by Florey and Chain and colleagues at Oxford University that was responsible for the discovery and development of the active chemicals produced by the organism.
40. White, T.J., Bruns, T., Lee, S., Taylor J., (1990). Amplification and direct sequencing of fungal ribosomal RNA genes for phylogenetics. In *PCR Protocols: A Guide to Methods and Applications* (M.A. Innis, D.H. Gelfand, J.J. Sninsky and T.J. White, Eds.): pp.315-322. Academic Press, San Diego, USA.
41. Zhenguo, W., Chen, P. and Xie, P. (Eds.) (1999). In *History and Development of Traditional Chinese Medicine*. Vol 1. Science Press: Beijing.

Psychrophilic bacteria: from detection to commercialization

J. P. Bowman

Australian Food Safety Centre of Excellence

School of Agricultural Science, University of Tasmania, Hobart, Tasmania, Australia

1. Psychrophilic bacteria and relevance to biodiscovery

Psychrophilic organisms are specialised for growth at relatively low temperatures, typically exhibiting optimal growth rates at 5-15°C. At temperatures at 25°C or more they rapidly die and thus they represent an extremophile with specific molecular and biochemical adaptations which are of interest to ecologists, molecular biologists, biochemists and biotechnologists. Though psychrophilic bacteria have been known to exist for several decades (e.g. reviewed in the seminal work of R.Y. Morita [24]), it is only in the last decade that a heightened interest in psychrophiles has occurred. This comes partly from the interest in extremophiles in general, particularly hyperthermophiles, in which various biotechnology (Table 1) and astrobiology oriented initatives have driven research and encouraged bioprospecting for more examples to investigate. In the case of psychrophilic bacteria, the research is also linked to studies of life in Polar Regions in which they proliferate. Psychrophilic bacteria do not occur in environments affected by intermittent heating (i.e. from solar heating) or aquatic ecosystems which have seasonal temperatures increases above 4°C. Marine ecosystems have been found to harbour by far the highest diversity of psychrophiles.

16S rRNA gene-based molecular surveys of cold ecosystems including sea-ice and polar and continental shelf sediments indicate psychrophilic bacteria occur in most known bacterial phyla but are absent from deeply branching lineages which contain predominantly thermophilic taxa. Indeed, psychrophilic taxa only occur at the tips of evolutionary lineages suggesting they have only recently evolved. Psychrophilic bacteria have mostly so far been cultivated from polar marine or marine-derived ecosystems, deep-sea trench sediments while terrestrial psychrophiles come almost entirely from Siberian tundra. Most new species belong to the *Proteobacteria* (mostly

Table 1: Antarctic and Southern Ocean ecosystems corresponding conditions and the possibilities for biotechnology

Habitat	Defining condition	Bioprospecting opportunities
Seawater, maritime lakes, sea-ice	Low temperature	Cold-active enzymes/ catalysts; bioremediation; surfactants; anti-freezes; polyunsaturated fatty acids; novel pigments
Southern Ocean seawater	Low nutrient concentration	High affinity catalysts and ligands
Hypersaline lakes	High salinity	Halotolerant enzymes; novel metabolites; novel pigments
Marine/lake sediment	Anaerobic/low temperature	Anaerobic biotransformations; novel bioactives
Soil, lithic habitats	Cold, dry	Novel bioactives

Gammaproteobacteria), *Bacteroidetes* and to a lesser extent the *Actinobacteria* (mostly suborder *Micrococcinaeae*).

The genetic potential of cold ecosystems is profound and their exploration has only just begun. For example, in a study of Southern Ocean surface waters, the *Bacteroidetes* were revealed to be surprisingly diverse and predominant in the cold Antarctic marine waters (annual temperature nearly a constant -1.8°C). Cultivation efforts so far have only scratched the surface of this group. In marine sediment all indigenous microflora is cold adapted and represents a nearly completely untapped biological pool of tremendous genetic and biotechnological potential and includes many large novels groups. Cold adapted *Archaea* represent evolutionary radiations radically distinct from the more well-known cultivated, mesophilic to hyperthermophilic lineages.

Cold adaptation appears species specific and unlike thermophily, psychrophily appears to involve specific and probably quite subtle alterations to cellular biochemistry which are not yet understood in any real detail, especially at the genetic level. Currently, complete or draft genomes are available for a number of psychrophilic prokaryotes, including: *Methanogenium frigidium*, *Desulfotalea psychrophila* and *Colwellia psychrerythraea* strain 34H and this wealth of genetic data may shed light on the molecular basis of cold adaptation.

Cold-type adapations that have been most studied, attracted patents and subsequently commercialized in some form include cold-active enzymes and polyunsaturated fatty acids. Cold active enzymes include proteins in which the physical and chemical structure has been altered to make them more flexible. Cold imposes rigidity on enzyme structures, reducing access of substrates to active sites and thus reducing or eliminating catalytic activity (i.e. cold denaturation). Several biochemical modifications of protein structures have been elucidated to avoid cold denaturation helped by the accumulation of knowledge from X-Ray crystallographic studies. Interestingly, enzymes with quite different temperature optima have been found to have identical amino acid sequences thus modification of proteins to make them cold-active may often occur at post-translational level. Cold active enzymes have several applications especially in the food industry, janitorial industries, synthesis of heat-labile chemicals and could be utilized to simply reduce energy useage. Specific examples in the food industry include the production of lactose-free milk (β-galactosidase) and tenderisation of meat (proteases) which through the use of cold-active enzymes reduces the risk of food spoilage and bacterial contamination. In janitorial industries the highest use would be in cold water detergents. Psychrophiles have been recently indicated as an alternative source of antifreeze proteins which can be used in the food industry (keeping frozen foods fresher and tastier) and in cryopreservation of tissues, organs, blood and other products with finite shelf life.

Several species of psychrophilic bacteria are also sources of polyunsaturated fatty acids (PUFA) which are well-documented nutriceuticals with health and economic benefits. These PUFAs include eicosapentaenoic, docosahexaenoic and arachidonic acids and are produced by psychrophiles to improve the fluidity of their cytoplasmic membranes which tend to become more rigid and thus less permeable (to nutrients) at low temperature. The bacteria also act as a primary de nova source of these lipids in ecosystems for organisms lacking the ability to synthesize them. Fish are currently a major source of nutriceutical PUFA but with environmental concerns about declining fish stocks alternative sources have been sought. Bacteria provide a pure source of PUFA (one or two major types) and since the enzyme systems responsible for the synthesis of PUFA is cytoplasmic, unlike membrane -associated enzymes in eukaryotes, fullscale industrial output is potentially more technologically straightforward. Genetic engineering of plant crops and other organisms are also possible using cloned bacterial PUFA synthesis gene operons. Applications shown in Table 2 have been reviewed extensively in the literature [e.g. 13, 28].

2. Microbial biodiversity in the Antarctic

The inherent lack of knowledge on the microbiota (including microscopic bacteria, plants and animals) of Antarctica and the Southern Ocean are attractive for biodiscovery and biotechnology-related research although this has recently attracted controversy (see below). The true interest that polar ecosystems provide for biotechnology are the environmental extremes found there - permanent cold, aridity of the ice-free zones on the continent, extremely salty and unusual maritime lakes etc.. These exotic locations contain novel diversity that has been extensively but far from completely surveyed in the last 5 years (Table 2). The diversity found has both a significant potential for basic and for applied science. A multiplicity of biodiscovery opportunities is clearly available for study (see below for caveats to this in regards to the Antarctic Treaty). Antarctic (and Arctic) microbial diversity has been investigated only relatively recently on a purely biotechnological level and outcomes advocated include novel sources of polyunsaturated fatty acids and cold-active enzymes (Figure 1), pigments, expolysaccharides and secondary metabolites [e.g. 13, 23, 28, 29].

The Antarctic Microbiology group at the University of Tasmania has been investigating prokaryotes since about the late 1980s. Originally, the focus was on microbial processes. Studies conducted by Peter Franzmann and colleagues investigated sulfur cycling and methane production as well as temperature adaptive responses [16]. A contract extant between 1995-2001 with a company now referred to as Cerylid Pty. Ltd. allowed the acquisition of a collection over >4000 strains including a large number of *Actinobacteria* and actinomycetes from Antarctic and sub-Antarctic terrestrial samples (mostly glacially deposited and ornithogenic soils). These strains were ostensibly targeted for pharmaceuticals. In 2004 these strains are being investigated for hydrocarbon degradative ability and anti-food pathogen secondary metabolites through an Australian Biological Resources project. About 30 new Antarctic species and 10 genera have been described (several are listed in 28, 29].

Figure 1: Resting cell serine protease activity (substrate hide powder azure) of various Antarctic sea-ice psychrophilic bacteria (*Shewanella gelidimarina* ACAM456; *Pseudoalteromonas antarctica* IC000 and *Colwellia hornerae* ACAM 607) over a range of temperatures. The wide temperature range of activity in some strains indicates multiple isozymes with different temperature are synthesized simultaneously. Data from A. Buia [4]

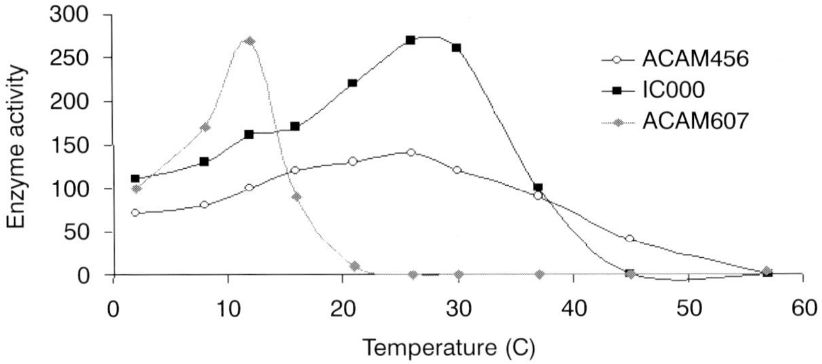

Table 2: Surveys and analyses of novel microbial (Bacteria and Archaea) diversity in Antarctic ecosystems

Habitats Investigated	Reference
Sea-ice	Bowman *et al.* 1997; Brown and Bowman 2001; Brinkmeyer *et al.* 2003
Glacial ice	Christner *et al.* 2003
Quartz stone subliths	Smith *et. al*. 2000
Antarctic lakes	Bowman *et al.* 2000a, 2000b; Gordon *et al.* 2000 Van Trappen *et al.* 2002; Taton *et al.* 2003; Karr *et al.* 2003.
Seawater	Lopez-Garcia *et al.* 2001; Bano *et al.* 2004; Abell and Bowman 2004
Marine sediment	Bowman and McCuaig 2003, Bowman *et al.* 2003, Powell *et al.* 2003, Purdy *et al.* 2003
Epibenthic fauna	Webster *et al.* 2004

More recently, in line with trends in microbial ecology, assessments of microbial diversity and biogeography within Antarctic ecosystems have been performed using purely molecular methods, e.g. denaturing gradient gel electrophoresis, fluorescent in situ hybridisation, clone libraries (used in various studies listed in Table 2). The primary outcomes of the molecular ecological studies provided determination of functional and structural relationships in particular ecosystems as well as observing the effects of temperature (especially cold) and other factors such as the effect of primary production on diversity and community structure. Springing from this work the scope for further explorations of polar microbial diversity is extremely broad and substantial effort will need to be of a targeted nature.

Two recent examples of microbial diversity in Antarctic ecosystems are given as examples here to emphasize the range of conditions possible in the Antarctic region and also to indicate the novel microbial diversity present which has yet to be properly accessed.

2.1 Marine sediment collected off Antarctic (Mertz Polynya)

Molecular studies have been recently performed on surface sediments collected from the continental shelf off of Eastern Antarctica. Using a variety of techniques the communities living in the sediment at the same depth appeared to be relatively homogenous (sampling over ~3000 km^2). Examining the depth profile of a single core using higher resolution methods however revealed extremely high prokaryote diversity [10] (Fig. 2) with the community changing significantly over sediment depth [9]. Diversity found was similar to that of soil and other sediments from temperate sites [32]. Almost all bacterial phyla not dominated by thermophilic taxa were detected (Fig.3). Archaeal diversity was also broad and totally divergent from current cultured Archaea which has a bias towards terrestrial and/or thermophilic species (Fig. 4). Studies comparing the Antarctic samples with temperate samples (collected off Tasmania) supported a hypothesis that psychrophiles are a recent development in Earth's evolution [5]. On the basis of the work the pool of genetic potential in marine sediment is incredibly enormous but so far has not been accessed at all. Techniques such as environmental genomes/metabolomes should provide a wealth of biodiscovery. An interesting and important fact (especially from a conservation point of view) is that it could be achieved from virtually *any* marine sediment sample, not necessarily Antarctic, including those outside of special and unique ecosystems.

Figure 2: Prokaryotic species richness in a single marine sediment core collected from a continental shelf site off Antartica.

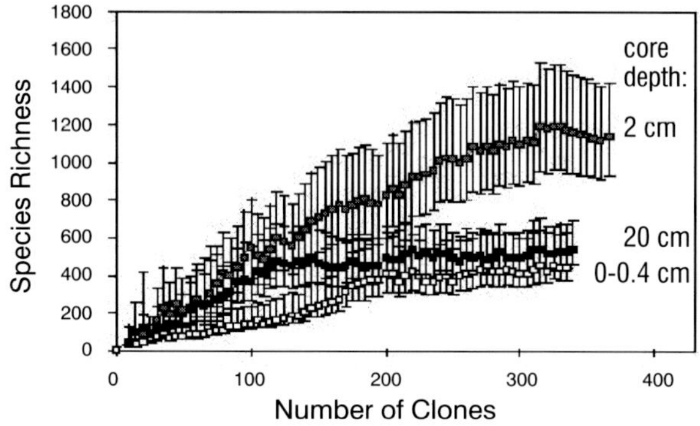

Data is based on 16S rRNA gene clone library analysis of three different sections of the core (depths 0-0.4 cm, 1-2 cm and 20-21 cm). Species are the equivalent of phylotypes, clone sequences with dissimilarities of <0.02. Integrated species richness of the three libraries was 4350 (± 380) indicating total diversity through the surface benthos is likely much higher. Data from Bowman and McCuaig [10].

Figure 3: 16S rRNA gene phylogenetic tree indicating the diverse spread of psychrophilic bacteria.

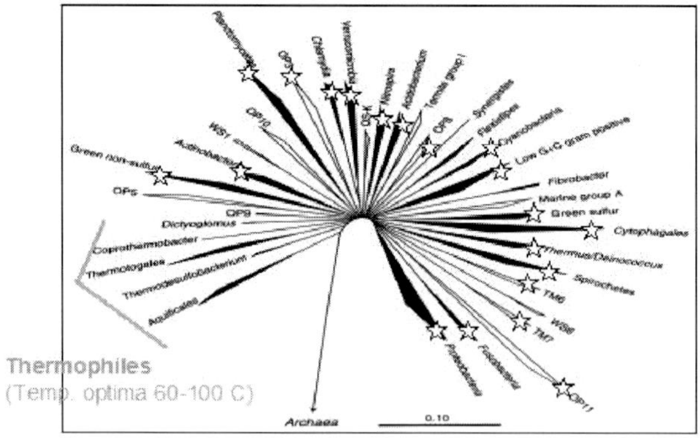

Stars denote the given phylum equivalent branch contains psychrophilic bacteria, either isolates or clones obtained through molecular surveys of permanently cold ecosystems. Psychrophilic taxa are absent from phyla dominated by thermophiles. Tree adapted from Hugenholtz *et al.* [19].

Figure 4: Archaeal diversity in cold sediment samples (black lines) differs dramatically from cultured, predominantly thermophilic, Archaea indicating substantial novel genetic diversity and novelty remains to be discovered.

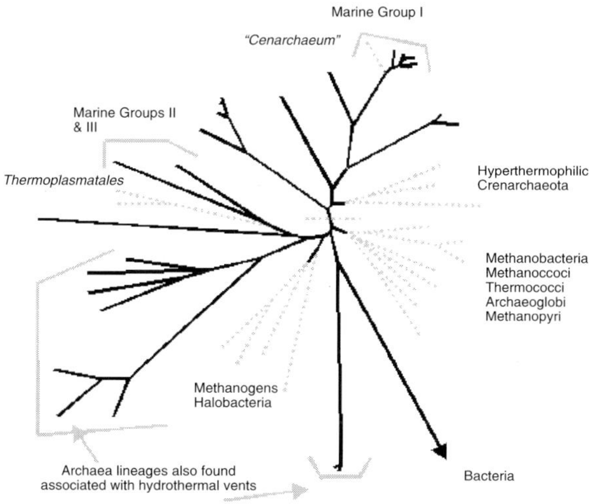

Data adapted from Bowman and McCuaig [10].

2.2. *Flavobacteria* in the Southern Ocean

The high diversity of the phylum "*Bacteroidetes*" (*Cytophaga-Flavobacterium-Bacteroides* group) has been noticed to be strongly correlated to primary production (algal photosynthesis and growth) [e.g. 11], indeed recent data suggest the diversity follows a "U-shaped" relationship [17] in which diversity peaks at some nominal point, a behaviour much like plants and animals. A high diversity of class *Flavobacteria* has been found in Southern Ocean samples (e.g. Fig. 5), particularly in the colder, relatively more productive waters of the Antarctic Zone [1]. Overall, the group has yet to be accessed for any sort of scientific investigation let alone biotechnology. As class *Flavobacteria* are believed to be involved heavily in the remineralization of organic matter in the oceans (thus the association with algae) it is likely the group represents a biotechnologically relevant bioresource that could be sampled without any real environmental impact. Seawater could thus act as ready source of biodiscovery using innovative culture techniques such as high throuput culturing [e.g. 15]. The taxonomy of the group has rapidly expanded in the last few years indicating the active interest in the group.

3. Biodiscovery and the Antarctic Treaty

The Antarctic Treaty was established in 1961 to protect Antarctica from uncontrolled commercial exploitation (e.g. mining), balkanization and militarization. Every 10 years or so, the Antarctic Treaty is updated. In 1991 the Madrid Protocol was established which contains several annexes pertaining to protection of the Antarctic from environmental degradation through human activity. More recently, several Antarctic and Southern Ocean Coalition (ASOC) members have noticed emerging trends towards increased commercialization which were considered disturbing and express the belief that they threaten to "*overwhelm the Antarctic Treaty's and the Protocol's core values of science, peace and cooperation*" (ACTM XXII). Indeed, some in the popular press end of the scientific and policy community see "bio-piracy" as a significant threat to Antarctica [e.g. 21].

Figure 5: The diversity of class *Flavobacteria* detected in surface Southern Ocean seawater samples using a specific 16S rRNA gene primer set is high in the regions surrounding Antarctica (water temperature -1.2 to –1.8°C).

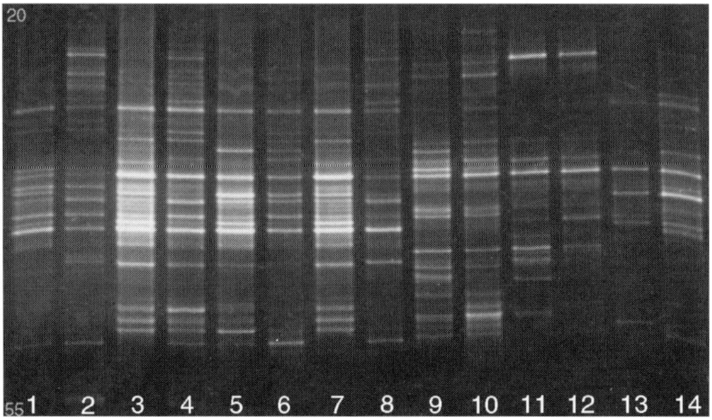

PCR products are resolved by DGGE (20-55% formamide/urea gradient, temperature constant at 60°C). Figure Diversity of Archaea found in Antarctic marine sediment indicated by the black lines. Data from Abell and Bowman [1].

The Antarctic Treaty advisory body, the Scientific Committee on Antarctic Research (SCAR), have also raised recently several concerns about bioprospecting quoting: "*While no current instance of harvesting for biotechnology is known, there are obvious environmental ramifications of the taking of animals and plants as a commercial venture*"....."*may develop into important pressures on Antarctic resources*". Overall, popular press suggests there is new "cold rush" in the Antarctic (and perhaps more quietly in the Arctic) e.g. *"On Thin Ice: How the Quest for a Billion-Dollar Microbe is Running Out of Control in Earth's Last Wilderness*" (www.commondreams.org). With these pronounced concerns a sensible, sensitive and highly justifiable approach must be conducted in the study of Antarctic microbial diversity for *any* form of commercial gain. It is quite obvious that strict new guidelines governing Antarctic biodiscovery will be introduced into a new iteration of the Antarctic Treaty in the very near future e.g.:

- Material/samples taken from Antarctic environment for biodiversity analysis should be minimized, in other words "bioprospecting" and therefore "biodiscovery" must be contained to a limited level and *must* be accompanied by quality fundamental science.
- Environmental impact assessments need to be completed (as handled by the appropriate coordinating national Antarctic science organization e.g. Australian Antarctic Division, British Antarctic Survey etc.) even if the impact would be negligible so that a database of "sampling" (where, when, how etc.) can be developed.
- Biotechnology based research should only be conducted on organisms derived from the original source material e.g. bacterial isolates, genomic DNA. Recollecting for only commercial reasons must be banned as under the existing Antarctic Treaty as it can be considered to be akin to the mining of minerals.
- Areas designated as Special Sites of Scientific Interest (SSSI) should be interdicted from any form of commercial oriented investigations.

This area will continue to be of future (and heated) discussion especially if biotechnology interests increase from studies of microbiota to larger forms of life such as inshore marine fauna and flora which are currently regulated under CCAMLR (Commission for the Conservation of Antarctic Marine Living Resources). Better understanding of microbial ecosystems and community members therein will provide more sensible outcomes for biotechnology which hopefully will be seen as useful and helpful not purely done for profit, exploitation and a need to justify science research dollars. In the case of the Antarctica and other sensitive and unique ecosystems, fundamental science should always take precedence since we still know so little about these places. Biotechnology (and commercialisation) can only benefit from a richer understanding of natural ecosystems and thus should be considered a goal of scientific labours, not *the* goal.

References

1. Abell, G.C.J. and Bowman, J.P. (2004). Ecological and biogeographic relationships of particle associated and planktonic *Flavobacteria* in the Southern Ocean. *FEMS Microbiology Ecology*, in press.
2. ACTM XXII. Antarctic and Southern Ocean Coalition report on the XXIIth Antarctic Treaty consultative meeting 1999. http://www.asoc.org/currentpress/XXIImeeting.htm
3. Bano, N., Ruffin, S., Ransom, B. and Hollibaugh, J.T. (2004). Phylogenetic composition of Arctic Ocean archaeal assemblages and comparison with Antarctic assemblages. *Applied and Environmental Microbiology,* 70: 781-789.
4. Buia, A. (1997). *Cold adapted enzyme activity in Antarctic sea-ice bacteria.* Honours Thesis. University of Tasmania.
5. Bowman, J.P. (2004). Psychrophilic prokaryote structural-functional relationships, biogeography and evolution within marine sediment. *Cellular and Molecular Biology,* in press.
6. Bowman, J.P., McCammon, S.A., Brown, M.V., Nichols, D.S. and McMeekin, T.A. (1997). Diversity and association of psychrophilic bacteria in Antarctic sea ice. *Applied and Environmental Microbiology,* **63**: 3068-3078.
7. Bowman, J.P., Rea, S.M., McCammon, S.A. and McMeekin, T.A. (2000). Diversity and community structure within anoxic sediment from marine salinity meromictic lakes and a coastal meromictic marine basin, Vestfold Hills, Eastern Antarctica. *Environmental Microbiology,* **2**: 227-237.
8. Bowman, J.P., McCammon, S.A., Rea, S.M. and McMeekin, T.A. (2000). The microbial composition of three limnologically disparate hypersaline Antarctic lakes. *FEMS Microbiology Letters,* **183**: 81-88.
9. Bowman J.P., McCammon, S.A., Gibson, J.A.E, Nichols, P.D. and Robertson, L. (2003). Microbial metabolic activity and community structure within Antarctic continental shelf sediment. *Applied and Environmental Microbiology,* **69**: 2448-2462.
10. Bowman, J.P. and McCuaig, R.D. (2003). Diversity and biogeography of prokaryotes dwelling in Antarctic continental shelf sediment. *Applied Environmental Microbiology,* **69**: 2463-2484.
11. Brinkmeyer, R., Knittel, K., Jurgens, J., Weyland, H., Amann, R. and Helmke. E. (2003). Diversity and structure of bacterial communities in Arctic versus Antarctic pack ice. *Applied and Environmental Microbiology,* **69**: 6610-6619.
12. Brown, M.V. and Bowman, J.P. (2001). A molecular phylogenetic survey of sea-ice microbial communities (SIMCO). *FEMS Microbiology Ecology,* **35**: 267-275.
13. Cavicchioli, R., Siddiqui, K.S., Andrews D. and Sowers KR (2002). Low-temperature extremophiles and their applications. *Current Opinions in Biotechnology,* **13**: 253-261.
14. Christner, B.C., Kvitko, B.H. and Reeve, J.N. (2003). Molecular identification of Bacteria and Eukarya inhabiting an Antarctic cryoconite hole. *Extremophiles,* **7**: 177-183.
15. Connon, S.A. and Giovannoni, S.J. (2002). High-throughput methods for culturing microorganisms in very-low-nutrient media yield diverse new marine isolates. *Applied and Environmental Microbiology,* **68**: 3878-3885.

16. Franzmann, P.D. (1996). Examination of Antarctic prokaryotic diversity through molecular comparisons. *Biodiversity and Conservation,* **5:** 1295-1305.

17. Gordon, D.A., Priscu, J. and Giovannoni, S. (2000). Origin and phylogeny of microbes living in permanent Antarctic lake ice. *Microbial Ecology,* **39**: 197-202.

18. Horner-Devine, C.M., Leibold, M.A., Smith, V.H. and Bohannan, B.J.M. (2003). Bacterial diversity patterns along a gradient of primary productivity. *Ecology Letters,* **6**: 613-622.

19. Hugenholtz, P., Goebel, B.M. and Pace, N.R. (1998). Impact of culture-independent studies on the emerging phylogenetic view of bacterial diversity. *Journal of Bacteriology,* **180:** 4765-4774.

20. Karr, E.A., Sattley, W.M., Jung. D.O., Madigan, M.T. and Achenbach, L.A. (2003). Remarkable Diversity of Phototrophic Purple Bacteria in a Permanently Frozen Antarctic Lake. *Applied and Environmental Microbiology,* **69**: 4910-4914.

21. Leahy, S. (2004). "Bio-Pirates of the Antarctic". ZNet. http://www.zmag.org/weluser.htm

22. Lopez-Garcia, P., Lopez-Lopez, A., Moreira, D. and Rodriguez-Valera, F. (2001). Diversity of free-living prokaryotes from a deep-sea site at the Antarctic Polar Front. *FEMS Microbiology Ecology,* **36**: 193-202.

23. Mancuso Nichols C.A., Bowman J.P., Geron S., Raguénès G. and Guezennec J. (2004). Exopolysaccharides from Antarctic bacteria. *Journal of Applied Microbiology,* **96:** 1057-1066.

24. Morita, R.Y. (1975). Psychrophilic bacteria. *Bacteriology Reviews,* **39**: 144-167.

25. Powell, S.M., Bowman, J.P., Snape, I. and Stark, J.S. (2003). Microbial community variation in pristine and polluted nearshore Antarctic sediments. *FEMS Microbiology Ecology,* **45**: 135-145.

26. Purdy, K.J., Nedwell, D.B. and Embley, T.M. (2003). Analysis of the sulfate-reducing bacterial and methanogenic archaeal populations in contrasting Antarctic sediments. *Applied and Environmental Microbiology*, **69**: 3181-3191.

27. Smith, M.C., Bowman, J.P., Scott, F.J. and Line, M.A. (2000). Sublithic bacteria associated with Antarctic quartz stones. *Antarctic Science,* **12**: 177-184.

28. Nichols, D.S., Bowman, J.P., Sanderson, K., Mancuso Nichols, C., Lewis, T., McMeekin, T.A. and Nichols, P.D. (1999). Developments with Antarctic microorganisms: culture collections, bioactivity screening, taxonomy, PUFA production and cold-adapted enzymes. *Current Opinions in Biotechnology,* **10**: 240-246.

29. Nichols, D.S., Sanderson, K., Buia, A., van de Kamp, J., Holloway, P., Bowman, J.P., Smith, M., Mancuso Nichols, C., Nichols, P.D. and McMeekin, T.A. (2001). Bioprospecting and biotechnology in Antarctica. *The Antarctic: Past, Present and Future* (Jabour-Green, J., Hayward, M. Eds.). Antarctic CRC Research Report 28, Antarctic CRC, Hobart, Australia, pp.85-105.

30. Reaka-Kudla, M.L., Wilson, D.E. and Wilson, E.O. (Eds.) (1997). *Biodiversity II: understanding and protecting our biological resources*. Joseph Henry Press, Washington, D.C.

31. Taton, A., Grubisic, S., Brambilla, E., De Wit, R., and Wilmotte, A. (2003). Cyanobacterial diversity in natural and artificial microbial mats of Lake Fryxell (McMurdo Dry Valleys, Antarctica): a morphological and molecular approach. *Applied and Environmental Microbiology,* **69**: 5157-5169.

32. Torsvik, V., Sorheim, R., and Goksoyr, J. (1996). Total bacterial diversity in soil and sediment communities - A review. *Journal of Industrial Microbiology,* **17**: 170-178.

33. Van Trappen, S., Mergaert, J., Van Eygen, S., Dawyndt, P., Cnockaert, M.C. and Swings, J. (2002). Diversity of 746 heterotrophic bacteria isolated from microbial mats from ten Antarctic lakes. *Systematic and Applied Microbiology*, **25**: 603-610.

34. Webster, N.S., Wilson, K.J., Blackall, L.L. and Hill, R.T. (2001). Phylogenetic Diversity of Bacteria Associated with the Marine Sponge *Rhopaloeides odorabile. Applied and Environmental Microbiology*, **67**: 434-444.

Biotechnological exploitation of heterotrophic bacteria and filamentous fungi isolated from benthic mats of Antarctic lakes

F. Marinelli[1], M. Brunati[1], F. Sponga[1], I. Ciciliato[1], D. Losi[1], S. Van Trappen[2], J. Mergaert[2], J. Swings[2], E. Göttlich[3], S. de Hoog[4], J. Luis Rojas[5] and O. Genilloud[5]

[1]Vicuron Pharmaceuticals (formerly Biosearch Italia S.p.A), Via R. Lepetit 34, 21040 Gerenzano, Varese, Italy;

[2]Laboratorium voor Microbiologie, Universiteit Gent, K. L. Ledeganckstraat 35, B-9000 Gent, Belgium;

[3]Rheinisch Westfalisches Institut für Wasserforschung, Moritz Str. 26, D-45476 Mülheim an der Ruhr, Germany;

[4]Centraalbureau voor Schimmelcultures, P.O. Box 85167, NL-3508 AD Utrecht, The Netherlands;

[5]CIBE, Merck Research Laboratories, Merck Sharp and Dohme de España S.A., Josefa Valcárcel 38, E-28027 Madrid, Spain

1. The MICROMAT project

Natural products have been a critically important source of clinically relevant therapeutic molecules. However, the discovery rate of novel structural classes of antimicrobial molecules has declined in the last decade [21]. Recent progress in molecular microbial ecology shows that the extent of microbial diversity in nature is far greater than previously thought, as the number of known species is less than 1 % [33]. A renewed interest in examining microorganisms for novel pharmaceuticals has stimulated the development of integrated approaches combining specific isolation methods and the access to geographically diverse sample sources and to different ecological niches [31]. Metabolic potential is also being exploited by cloning microbial genes in environmental libraries without undergoing the step of culturing microbes [21].

In this context, the academic and industrial joint research of the EC project MICROMAT (http://www.nerc-bas.ac.uk/public/mlsd/micromat) has focused on the study of the culturable and uncultivable – or, better, the "yet uncultivable" – diversity in microbial mats growing in Antarctic lakes. The Antarctic continent and its surrounding marine sites offer a unique opportunity to investigate an unexplored microbial biodiversity [4, 8]. Antarctica is in fact characterized by its geographical and climatic isolation. The extreme climate has led to evolution of novel biochemical adaptations to severe low temperatures and the possibility of indigenous species. Moreover most of the continent has experienced little or no anthropogenic influence. Antarctic lakes include both freshwater and hyper-saline systems and some of them are covered by perennial ice [13, 49]. Their benthic areas receive sufficient solar radiation to be covered by microbial mats. These benthic mats have accumulated for thousands of years and are virtually undisturbed due to the particular climatic conditions and the absence of higher metazoans. Results from the MICROMAT project have pointed out the extremely high microbial diversity in mats where numerous novel phylotypes have and are being described [8, 40, 42, 43 and unpublished results).

1.1 Sampling

Mats were sampled from 13 lakes of different ages and physico-chemical characteristics located in three distinct regions of the Antarctic continent (Fig.1). The Larsemann Hills are a series of granite and gneiss peninsulas into Prydz Bay (Eastern Antarctica) with fjords and lakes directly (currents, inlets) or indirectly (sea spray) subjected to marine influences; most of them thaw for up to 2 months in summer and during this time are subjected to considerable wind driven mixing. The Vestfold Hills constitute a low-lying area situated South of the Larsemann Hills, where hundreds of water bodies are found in the valleys, with salinities ranging from fresh to hypersaline (ten times seawater) [7]. In contrast to the Larsemann and Vestfold Hills, the lakes in the McMurdo Dry Valleys of South Victoria Land are very old (hundreds of thousand of years). They do not lose their ice-cover and thus lack any turbulence. They vary from dilute meltwaters to hypersaline lakes [13, 19, 49].

Samples of the mats were taken either at a water depth of 3-4 m or from the littoral zone of the lakes [42]. The diversity of bacteria, cyanobacteria, algae, protozoans and fungi in the microbial mat samples of several Antarctic lakes was studied by conventional (direct microscopy, cultivation) and molecular methods (clone libraries and DGGE based on the SSU rDNA from the samples).

Figure 1: Three distinct regions of the Antarctic continent

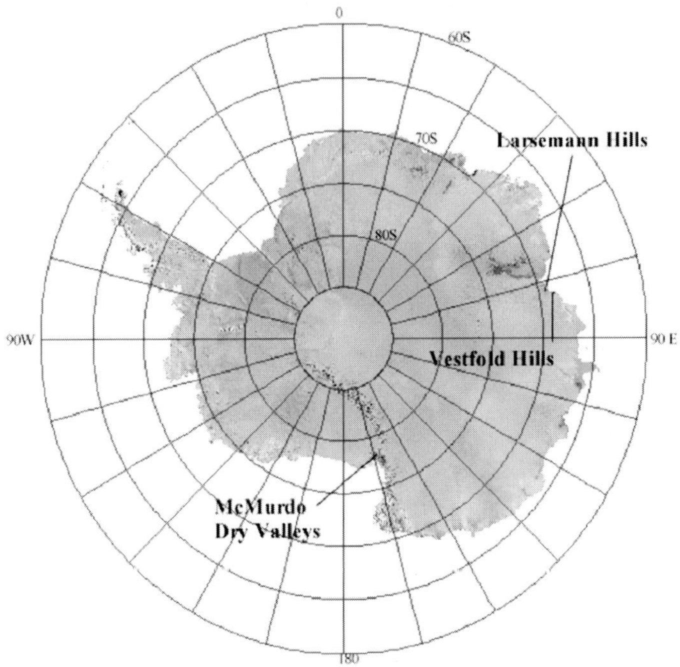

More than 1500 strains were isolated and a part of them was screened for the production of new cold-adapted enzymes and antimicrobial compounds. In this report we describe our combined efforts to screen 158 filamentous fungi and 723 heterotrophic bacteria isolated from mats collected in five lakes in the Larsemann Hills, six in the Vestfold Hills and two in the McMurdo Dry Valleys. Our main objective was to assess the potential of this microbial diversity with the aim of discovering new anti-infective producers among these Antarctic microorganisms.

1.2 Diversity of Antarctic isolates

Table 1 and Table 2 show the geographical distribution and the lake of origin of the bacteria and fungi isolated from thirteen lakes located in the three distinct Antarctic regions sampled, that is, Larsemann Hills, Vestfold Hills and McMurdo Dry Valley. Taxonomical diversity among 746 Antarctic bacterial isolates, including the strains (723) tested in this work, was previously studied by fatty acid clustering analysis and by 16S rDNA sequencing of cluster representative strains: a dendrogram with

Table 1: Diversity and distribution of bacterial taxa isolated in different Antarctic areas

FAA cluster [a)]	Number of isolates tested [b)]	Phylogenetic branch [c)]	Nearest phylogenetic neighbour [d)]	Lake of origin [e)]	Area [f)]
1	1	CFB-branch	*Flavobacterium frigidarium* (98.0%)	GR	VH
2	5	CFB-branch	*Flavobacterium aquatile*, *F. tegetincola* (95.7%)	HI	VH
3	1	Not determined	Not determined	RE	LH
4	3	CFB-branch	*Saligentibacter salegens* (92.9%)	FR	DV
5	69	CFB-branch	*Flavobacterium gillisiae* (98.4%), *F. hydatis* (97.3%), *F. frigidarium* (97.5%)	FR (21), HO (2) PE (7), GR (17), WA (12), AC (1), DR (3), HI (1) RE (5)	DV VH LH
6	5	CFB-branch	*Flavobacterium gillisiae* (98.4%)	PE	VH
7	4	CFB-branch	*Flavobacterium hydatis* (96.9%)	FR (2) WA (2)	DV VH
8	1	Not determined	Not determined	FR	DV
9	2	CFB-branch	*Flavobacterium gillisiae* (94.6%)	HO (1), FR (1)	DV
10	25	CFB-branch	*Flavobacterium flevense* (95.3%)	AC (12), PE (4), WA (3), GR (2), DR (1) RE (3)	VH LH
11	4	CFB-branch	*Flavobacterium tegetincola* (98.2%)	AC (1), WA (2) RE (1)	VH LH
12	1	CFB-branch	*Flavobacterium tegetincola* (95.7%)	FR	DV
13	4	CFB-branch	*Gelidibacter algens* (99.8%)	FR (1) PE (1), AC (2)	DV VH
14	6	CFB-branch	*Hymenobacter actinosclerus* (97.3%)	FR (4) PE (1) RE (1)	DV VH LH
15	7	CFB-branch	*Cyclobacterium marinum* 92.5%)	FR (1) AC (2) RE (4)	DV VH LH
16	9	Gram + (high %GC)	*Microbacterium keratanolyticum* (96.0%)	DR (1), GR (1) RE (7)	VH LH

Table 1 continued:

FAA cluster [a)]	Number of isolates tested [b)]	Phylogenetic branch [c)]	Nearest phylogenetic neighbour [d)]	Lake of origin [e)]	Area [f)]
1	1	CFB-branch	*Flavobacterium frigidarium* (98.0%)	GR	VH
2	5	CFB-branch	*Flavobacterium aquatile, F. tegetincola* (95.7%)	HI	VH
3	1	Not determined	Not determined	RE	LH
4	3	CFB-branch	*Saligentibacter salegens* (92.9%)	FR	DV
5	69	CFB-branch	*Flavobacterium gillisiae* (98.4%), *F. hydatis* (97.3%), *F. frigidarium* (97.5%)	FR (21), HO (2) PE (7), GR (17), WA (12), AC (1), DR (3), HI (1) RE (5)	DV VH LH
6	5	CFB-branch	*Flavobacterium gillisiae* (98.4%)	PE	VH
7	4	CFB-branch	*Flavobacterium hydatis* (96.9%)	FR (2) WA (2)	DV VH
8	1	Not determined	Not determined	FR	DV
9	2	CFB-branch	*Flavobacterium gillisiae* (94.6%)	HO (1), FR (1)	DV
10	25	CFB-branch	*Flavobacterium flevense* (95.3%)	AC (12), PE (4), WA (3), GR (2), DR (1) RE (3)	VH LH
11	4	CFB-branch	*Flavobacterium tegetincola* (98.2%)	AC (1), WA (2) RE (1)	VH LH
12	1	CFB-branch	*Flavobacterium tegetincola* (95.7%)	FR	DV
13	4	CFB-branch	*Gelidibacter algens* (99.8%)	FR (1) PE (1), AC (2)	DV VH
14	6	CFB-branch	*Hymenobacter actinosclerus* (97.3%)	FR (4) PE (1) RE (1)	DV VH LH
15	7	CFB-branch	*Cyclobacterium marinum* (92.5%)	FR (1) AC (2) RE (4)	DV VH LH
16	9	Gram + (high %GC)	*Microbacterium keratanolyticum* (96.0%)	DR (1), GR (1) RE (7)	VH LH

Table 1 continued:

FAA cluster [a)]	Number of isolates tested [b)]	Phylogenetic branch [c)]	Nearest phylogenetic neighbour [d)]	Lake of origin [e)]	Area [f)]
32	2	Not determined	Not determined	AC	VH
33	5	α-*Proteobacteria*	*Porphyrobacter neustonensis* (97.7%)	AC (3) RE (2)	VH LH
34	3	α-*Proteobacteria*	*Porphyrobacter neustonensis* (97.6%)	AC (1) RE (2)	VH LH
35	5	α-*Proteobacteria*	*Devosia riboflavina* (97.2)	RE	LH
36	8	α-*Proteobacteria*	*Sphi ngomonas natatoria* (94.5%)	AC (1) RE (7)	VH LH
37	3	Not determined	Not determined	AC	VH
38	43	γ-*Proteobacteria*	*P sychrobacter glacincola* (97.5%)	FR (15), HO (1) AC (3), OR (1), PE (5), HI (3) RE (15)	DV VH LH
39	18	γ-*Proteobacteria*	*Marinobact er hydrocarbonoclasticus* (95.6%)	AC (7), OR (10), PE (1)	VH
40	16	α-*Proteobacteria*	*Brevundimonas subvibrioides* (99.1%)	FR (2 DR (3), OR (1), PE (1))RE (9)	DV VH LH
41	52	α-*Proteobacteria*	*Mesorhizobium loti* (96.9%), *Sulfitobacter pontiacus* (93.5%)	FR (10) AC (15), OR (15), PE (4), WA (1), HI (2) RE (5)	DV VH LH
NC	28	Not determined	Not determined	FR (8) OR (2), PE (3), GR (1), AC (8), HI (1) RE (5)	DV VH LH
Total	675				

a) The 41 fatty acid clusters were delineated in Van Trappen *et al.* [42]. 28 of the tested strains (NC means not clustered) formed single branches.

b) Data of cluster affiliation are reported on 675 bacterial strains out of the 723 screened in this paper. The fatty acid profiles of the remaining 48 strains -not listed in this table- were not clearly interpretable.

c) CFB branch means: *Cytophaga-Flavobacterium-Bacteroides* branch

d) The nearest phylogenetic neighbour was identified as described in Van Trappen et al [42]; (%) indicates the 16S rDNA sequence similarities based on pairwise alignments.

e) Numbers of strains isolated from each lake are indicated in parenthesis. The abbreviations FR, HO, RE, AC, DR, GR, HI, PE, WAand OR stand for lakes Fryxell, Hoare, Reid, Ace, Druzby, Grace, Highway, Pendant, Watts and Organic.

f) The abbreviations LH, VH, DV stand for Areas Larsemann Hills, Vestfold Hills and McMurdo Dry Valleys.

41 different fatty acid clusters and 31 strains forming single branches was described by Van Trappen *et al.* [42]. Table 1 reports the taxonomical distribution for 675 of 723 bacterial strains tested in this paper. The 675 isolates were distributed in all the 41 different fatty acid clusters previously identified and they also included 28 strains forming single branches. Fatty acid profiles on the rest of the screened strains (48 out of 723, not listed in Table 1) were not interpretable due to their low resolution or their strain cultivation conditions were unsuitable for general comparison (data not shown). The clustered bacteria were phylogenetically affiliated on the basis of their 16S rDNA sequences with several lineages in the alpha, beta and gamma subclasses of the *Proteobacteria*, the *Cytophaga-Flavobacterium-Bacteroides* branch, and the high and low percentage G+C Gram-positives. As shown in Table 1, most clusters (28) contained strains (614) isolated from different lakes, and often from different regions, suggesting that taxa showing these fatty acid compositions might be ubiquitous in Antarctic lakes.

The diversity of fungal Antarctic strains was determined on the basis of the taxonomical description provided by Göttlich *et al.* [15]. Table 2 shows the taxonomic distribution of fungi isolated from the three geographic areas. *Thelebolus* sp. was the predominant species in all the three regions sampled. *Thelebolus* was almost the only genus recovered from the old and isolated meromictic lakes of McMurdo Dry Valleys. Members of the most widely distributed genera such as *Penicillium*, *Aspergillus*, *Phoma*, *Cladosporium, Curvularia* were isolated in the Larsemann and Vestfold Hills areas. In terms of distribution of the isolated fungi among different genera, lakes in the Vestfold Hills turned out to be the most productive, as some representatives of genera *Botrytis*, *Beauveria*, *Acremonium*, *Arthrinium*, *Phialophora, Geomyces, Alternaria* were isolated only from these sources.

1.3 Cultivation of Antarctic isolates

The cultivation of Antarctic heterotrophic bacteria and filamentous fungi was optimized in terms of medium composition and incubation temperature by a screening of different nutrients in pre-culture and fermentative media [3, 5, 10, 12, 17, 23, 36] and by varying growth and production temperatures (unpublished data). Major changes in the fermentation conditions were the introduction of less concentrated media and the addition of marine salts, which are common adaptations of classical medium composition to the cultivation of marine microbes [4, 29, 30, 34]. About half of the bacteria grew better in marine derived media whereas for the other half rich classical media were suitable. In any cultivation condition used, all

Table 2: Diversity and distribution of fungal taxa isolated in the different Antarctic areas

Number of isolates tested	Order	Genus	Lake of origin[a)]	Area[b)]
71	*Leotiales*	*Thelebolus*	AC (7), DR (6), HI (4), WA (1) FR (9), HO (2) MA (5), OR (17), RE (20)	VH DV LH
2	*Eurotiales*	*Aspergillus*	PE (1) ST (1)	VH LH
16	*Eurotiales*	*Penicillium*	PE (12), WA (1), HI (1), AC (1) OR (1)	LH VH
1	*Helotiales*	*Botrytis*	PE	VH
3	*Clavicipitales*	*Beauveria*	PE (1), WA (2)	VH
1	*Hypocreales*	*Acremonium*	PE	VH
1	*Mitosporic Ascomycotina*	*Arthrinium*	PE	VH
17	*Dothideales*	*Cladosporium*	PE (8), AC (1) RE (6), MA (2)	VH LH
2	*Chaetothyriales*	*Phialophora*	WA	VH
9	*Dothideales*	*Phoma*	DR (3) MA (3), RE (3)	VH LH
4	*Onygenales*	*Geomyces*	PE (2), AC (2)	VH
1	*Pleosporales*	*Alternaria*	PE	VH
5	*Pleosporales*	*Curvularia*	PE (1) MA (4)	VH LH
1	*Pleosporales*	*Embellisia*	FR	DV
1	*Mitosporic Asmycotina*	*Onychophora*	FR	DR
23		non-identified	PE (4), HI (2), AC (1), WA (4) RE (15)	VH LH
Total 158				

a) Numbers of strains isolated from each lake were indicated in parenthesis. The abbreviations FR, HO, RE, AC, DR, HI, PE, WA, MA, OR and ST stand for lakes Fryxell, Hoare, Reid, Ace, Druzby, Highway, Pendant, Watts, Manning, Organic and Sarah Tarn

b) The abbreviations LH, VH, DV stand for Areas Larsemann Hills, Vestfold Hills and McMurdo Dry Valleys.

the bacteria and fungi produced a high biomass. When growth was monitored at different temperatures (4°C, 10°C, 22°C, 28°C), optimal temperatures were in the range of 22 to 28°C, suggesting that all these isolates are psychrotolerant rather than psychrophilic. However, *Thelebolus* strains produced more biomass around 20°C but sporulated at 10°C, indicating true psychrophily.

2. Antimicrobial activity

A total of 6,348 and 1,422 samples were prepared from the fermentation of the 723 bacteria and 158 fungi, respectively. Up to four different sample preparation methods - three based on the extraction by/with solvents with different polarity and another one consisting in the adsorption/elution to polystyrenic resin - were used to widen/favour the recovery of secondary metabolites with different molecular weights and lipo/hydrophilic properties/polarities. These samples were tested for growth inhibition against a panel of human pathogen microorganisms (*Staphylococcus aureus*, *Enterococcus faecium*, *Escherichia coli*, *Candida albicans*, *Aspergillus fumigatus* and *Cryptococcus neoformans*) either in solid or liquid assay formats.

Table 3 reports the frequency of microbiological activity against each test pathogen among the heterotrophic bacteria. From 124 bacteria out of 723 that were active, 110 produced only antibacterial activities and 14 showed some inhibition of the tested fungal strains. The frequency of antibacterial activity was normal against the Gram-positive *S. aureus* (13.2%) and high against the Gram-negative *E. coli* (7.4%). The active strains were distributed among 21 clusters out of the 41 above described and in several lineages in the alpha, beta and gamma subclasses of the *Proteobacteria*, the *Cytophaga-Flavobacterium-Bacteroides* branch, and the high and low percentage G+C Gram-positives. Almost 64 % of the active bacteria (79 isolates) were members of the 6 major fatty acid clusters (5, 22, 23, 26, 30, 41) that contained isolates widely distributed in the different sampling areas (see Table 1). In contrast, 39 active cultures were associated to 15 minor clusters; 4 strains formed single branches and 2 belonged to the not clustered group.

Further studies were performed with a subset of these strains (hits) that exhibited some relatively potent antibacterial activities against bacterial human pathogens (Table 4). None of the selected strains showed antifungal activity in the production conditions tested (data not shown). Two isolates (R-7513 and R-7941) obtained in the McMurdo Dry Valleys, specifically in the lakes Fryxell and Hoare, were studied more in detail. The isolates corresponded to coccoid high %GC Gram-positives,

Table 3: Distribution of isolates producing antimicrobial activities among bacterial taxa

Phylogenetic branch [a)]	FAA Cluster [b)]	Nearest phylogenetic neighbour [c)]	Strains tested [d)]	Active Strains	N° of Strains producing antibacterial activities on:			N°of Strains producing antifungal activities on:		
					S. aureus	*E. faecium*	*E. coli*	*C. albicans*	*C. neoformans*	*A. fumigatus*
CFB-branch	5	*Flavobacterium gillisiae* (98.4%), *F. hydatis* (97.3%), *F. frigidarium* (97.5%)	69	17	13	2	12	0	0	1
CFB-branch	7	*Flavobacterium hydatis* (96.9%)	4	2	2	0	1	0	0	0
CFB-branch	9	*Flavobacterium gillisiae* (94.6%)	2	1	1	1	0	0	0	0
CFB-branch	10	*Flavobacterium flevense* (95.3%)	25	4	3	1	1	0	0	0
CFB-branch	14	*Hymenobacter actinosclerus* (97.3%)	6	2	2	1	0	0	0	1
Gram + (high %GC)	16	*Microbacterium keratanolyticum* (96.0%)	9	1	1	0	0	0	0	0
Gram + (high %GC)	17	*Micrococcus luteus* (99.7%)	5	1	1	0	0	0	0	0
Gram + (high %GC)	18	*Arthrobacter agilis* (99.6%)	64	5	3	3	3	0	0	0
Gram + (low %GC)	20	*Bacillus oleronius* (93.2%)	12	3	1	1	3	0	0	0
α-*Proteobacteria*	40	*Brevundimonas subvibrioides* (99.1%)	16	3	3	0	2	0	0	0
α-*Proteobacteria*	41	*Mesorhizobium loti* (96.9%), *Sulfitobacter pontiacus* (93.5%)	52	9	9	2	5	0	0	0

a) CFC branch means: *Cytophaga-Flavobacterium-Bacteroides* branch

b) The fatty acid clusters were delineated in Van Trappen *et al.* [42]. In this table, only those fatty acid clusters comprising active isolates are listed, i.e. 21 out of the 41 previously described. 28 of the tested strains (NC) formed single branches. The fatty acid profiles of the remaining 48 strains were not clearly interpretable (NI). See also Table 1

Table 3: Continued

Phylogenetic branch [a)]	FAA Cluster [b)]	Nearest phylogenetic neighbour [c)]	Strains tested [d)]	Active Strains	N° of Strains producing antibacterial activities on:			N°of Strains producing antifungal activities on:		
β-*Proteobacteria*	24	*Pseudomonas saccharophila* (97.2%)	2	2	2	0	2	0	0	0
β-*Proteobacteria*	28	*Janthinobacterium lividum* (99.7%)	13	6	6	1	0	1	5	0
β-*Proteobacteria*	29	*Hydrogenophaga palleronii* (98.4%), *Aquaspirillum delicatum* (97.8%)	3	1	1	0	0	0	0	0
β-*Proteobacteria*	30	*Hydrogenophaga palleronii* (97.7%), *Janthinobacterium lividum* (99.6%)	42	11	11	1	3	1	1	1
γ-*Proteobacteria*	22	*Shewanella baltica* (98.7%)	74	19	17	0	7	0	2	0
γ-*Proteobacteria*	23	*Pseudomonas anguilliseptica, P.migulae* (95.8%)	74	10	5	1	8	0	0	0
γ-*Proteobacteria*	26	*Pseudomonas orientalis* (99.5%)	37	13	6	7	7	0	0	3
γ-*Proteobacteria*	38	*Psychrobacter glacincola* (97.5%)	43	5	2	3	4	0	0	0
γ-*Proteobacteria*	39	*Marinobacter hydrocarbonoclasticus* (95.6%)	18	2	2	0	0	0	0	0
Not determined	8	Not determined	1	1	1	0	0	0	0	0
Not determined	NC	Not determined	28	4	3	0	2	0	0	0
Not determined	NI	Not determined	48	2	1	0	1	0	0	0
		Total	647	124	96	24	61	2	8	6

c) The nearest phylogenetic neighbour was identified as described in Van Trappen *et al* [42] (%) indicates the 16S rDNA sequence similarities based on pairwise alignments.

d) Total strains tested per each of the 21clusters comprising active isolates.

which exhibited antibacterial activity against *B. subtilis*, *S. aureus* and *E. faecium*. Both strains, although isolated in different lakes, turned out to have similar fatty acid composition and 99.6 % 16SrDNA sequence similarity to *Arthrobacter agilis*. They apparently produced similar compounds. Preliminary mass spectrometry of purified extracts suggests these compounds to be cyclic thiazolyl peptide antibiotics with similar antibacterial spectrum of activity against Gram-positive bacteria (Z. Guan, personal communication). The other isolate (R-7687) from the Lake Hoare in McMurdo Dry Valleys belonged to the beta-proteobacteria and differs from the previous ones for the activity on *E. coli*. Seven other isolates, six from Lake Reid and one from Pendant, were also particularly active against Gram-positive and Gram-negative pathogens. Three of them are similar to *Pseudomonas orientalis,* whereas the others belonged to taxa isolated from cold, aquatic environments such as *Shewanella baltica* and *Psychrobacter glacincola.*

Table 5 shows the distribution of the antimicrobial activities among filamentous fungi belonging to different species, genera and orders. The frequency of antimicrobial activities was more than 29 % in total. As in the case of bacteria, the frequency of antibacterial activity was high against the Gram-positive *S. aureus* (14 %) and the Gram-negative *E. coli* (10 %). In contrast to bacteria, high antifungal activities against *C. albicans, A. fumigatus* and *C. neoformans* were obtained from these fungal isolates. It is worth noting that although *Thelebolus* was the taxon most frequently isolated (71 strains tested isolated from nine lakes in the three different areas), it was one of the less productive (only 11 active isolates, 15 %) in contrast to *Penicillium* and *Cladosporium*, two of the most active groups that were isolated from the Vestfold Hills and Larsemann Hills (see Table 2). *Penicillium* spp. were among the most active species with a rate of 93 % active strains, while lower percentages were obtained with *Cladosporium* spp. (35 %), two genera where comparable numbers of strains were tested.

Since many fungi are good producers of toxins, a cytotoxicity test based on the inhibition of labeled thymidine uptake in HeLa was introduced to select molecules not toxic to mammalian cells among those active extracts coming from fungi. Two thirds of the fungal extracts showed a high cytotoxicity against eukaryotic cells (data not shown). Table 6 reports the cytotoxicity test results and the antimicrobial spectrum of activity (expressed as end point inhibition) for eight selected fungi (hits). *Aspergillus clavatus* (IWW447) showed potent antibacterial and antifungal properties associated with a marked cytotoxic effect. This spectrum is typical of toxins produced by aspergilli

Table 4: Origin, taxonomical affiliation and antibacterial activities of selected bacteria isolates (Hits)

Area	Lake	Strain Code [a]	FAA Cluster	Phylogenetic branch	Nearest phylogenetic neighbour	Antibacterial activities in agar diffusion assays (mm) on: [b]		
						S. aureus	*E. faecium*	*E. coli*
Dry Valleys	Fryxell	R-7513	18	Gram + (high %GC)	*Arthrobacter agilis* (99.6%)	10	11	0
Dry Valleys	Hoare	R-7687	28	β-*Proteobacteria*	*Janthinobacterium lividum* (99.7%)	7	8	6
Dry Valleys	Hoare	R-7941	18	Gram + (high %GC)	*Arthrobacter agilis* (99.6%)	7	8	0
Vestfold Hills	Pendant	R-8990	22	γ-*Proteobacteria*	*Shewanella baltica* (98.7%)	7	0	7
Larsemann Hills	Reid	R-12565	26	γ-*Proteobacteria*	*Pseudomonas orientalis* (99.5%)	7	6	7
Larsemann Hills	Reid	R-12533	26	γ-*Proteobacteria*	*Pseudomonas orientalis* (99.5%)	6	8	6
Larsemann Hills	Reid	R-12535	26	γ-*Proteobacteria*	*Pseudomonas orientalis* (99.5%)	5	0	8
Larsemann Hills	Reid	R-12605	22	γ-*Proteobacteria*	*Shewanella baltica* (98.7%)	7	0	9
Larsemann Hills	Reid	R-12597	38	γ-*Proteobacteria*	*Psychrobacter glacincola* (97.5%)	5	0	6
Larsemann Hills	Reid	R-12583	NC	Not determined	Not determined	9	5	9

a) As preserved in the research collection of the Laboratory of Microbiology, University of Ghent, Belgium

b) Antibacterial activities were indicated as the diameter in mm of the inhibition zone obtained in agar diffusion assays.

Table 5: Distribution of isolates producing antimicrobial activities among fungal taxa

Order	Genus & species	Strains tested	Active strains	N° of Strains producing antibacterial activities on:			N° of Strains producing antifungal activities on:		
				S. aureus	*E. faecium*	*E. coli*	*C. albicans*	*C. neoformans*	*A. fumigatus*
Leotiales	*Thelebolus* sp.	71	11	6	0	0	5	4	1
Eurotiales	*Aspergillus clavatus*	1	1	1	1	1	1	1	0
Eurotiales	*Aspergillus niger*	1	0	0	0	0	0	0	0
Eurotiales	*Penicillium chrysogenum*	6	6	6	0	1	0	0	1
Eurotiales	*Penicillium crustosum*	2	2	0	0	1	1	0	0
Eurotiales	*Penicillium rugulosum*	1	1	1	1	1	0	0	0
Eurotiales	*Penicillium* sp.	7	6	1	0	5	3	3	0
Helotiales	*Botrytis* sp.	1	1	0	0	1	0	0	0
Clavicipitales	*Beauveria* sp.	3	2	0	0	0	1	1	1
Hypocreales	*Acremonium* sp.	1	0	0	0	0	0	0	0
Mitosporic Ascomycotina	*Arthrinium* sp.	1	0	0	0	0	0	0	0
Dothideales	*Cladosporium* sp.	8	6	4	0	2	1	1	0
Dothideales	*Cladosporium herbarum*	9	0	0	0	0	0	0	0
Chaetothyriales	*Phialophora* sp.	2	0	0	0	0	0	0	0
Dothideales	*Phoma* sp.	9	3	3	0	2	0	1	0
Onygenales	*Geomyces pannorum*	3	1	0	0	0	1	1	0
Onygenales	*Geomyces* sp.	1	0	0	0	0	0	0	0
Pleosporales	*Alternaria* sp.	1	0	0	0	0	0	0	0
Pleosporales	*Curvularia* sp.	5	1	0	0	0	1	0	0
Pleosporales	*Embellisia* sp.	1	0	0	0	0	0	0	0
Mitosporic Asmycotina	*Onychophora* sp.	1	0	0	0	0	0	0	0
	Non identified	23	6	0	0	2	3	0	4
	Total	158	47	22	2	16	17	12	7
	%		29	14	1	10	11	8	4

Table 6: Origin, taxonomical affiliation, cytotoxicity, antimicrobial activities of selected fungal isolates (Hits)

Area	Lake	Strain Code[a)]	Strain	Cyto-toxicity test [b)]	Antibacterial activities in liquid microtiter plate assay on: [c)]			Antifungal activities on liquid microtiter plate assay on: [c)]		
					S. aureus	*E. faecium*	*E. coli*	*C. albicans*	*C. neoformans*	*A. fumigatus*
Larsemann Hills	Sarah Tarn	IWW447	*Aspergillus clavatus*	160	2048	512	1024	128	2048	n.d
Vestfold Hills	Highway	IWW1035	*Penicillium chrysogenum*	10	32	8	8192	0	0	0
Vestfold Hills	Watts	IWW1023	*Penicillium crustosum*	10	0	0	16	0	0	0
Vestfold Hills	Pendant	IWW1054	*Penicillium* sp.	10	0	0	8	8	16	0
Vestfold Hills	Pendant	IWW1055	*Penicillium* sp.	20	32	0	8	0	0	16
Vestfold Hills	Ace	IWW1059	*Penicillium* sp.	20	0	0	8	16	16	0
Vestfold Hills	Pendant	IWW1017	*Beauveria* sp.	<10	0	0	0	0	8	0
Vestfold Hills	Pendant	IWW1019	*Cladosporium* sp.	<10	0	0	16	0	0	0

a) As preserved in the IWW Rheinisch Westfalisches Institut fur Wasserforschung Gemeinnutzige

b) Cytotoxicity is measured as endpoint in microdilution method, i.e. the highest dilution which inhibits 40% of HeLa cell thymidine uptake.

c) Antibacterial and antifungal activities are measured as endpoints in microdilution method, i.e. the highest dilution which inhibits 80% of test strain growth.

and other fungi. The five selected *Penicillium* strains showed different antimicrobial profiles associated with lower values in the cytotoxicity test, demonstrating that although isolated in the same region, they may produce diverse metabolic compounds. More interesting for the discovery of new specifically acting anti-infectives are those metabolites showing lower but more specific and not cytotoxic activities, such as the antifungal characteristics of *Beauveria* (IWW1017) or the specific *E. coli* inhibition effect of *Cladosporium* extract (IWW1019).

3. Discussion and general considerations

Although Antarctic lakes have been for a long time considered as inhospitable and isolated environments inhabited by species-poor communities, studies funded by the EC project MICROMAT revealed the presence of a large microbial diversity, especially for prokaryotes. Results from this interdisciplinary and polyphasic approach, of which part has been published up to now [8, 15, 40, 42, 44] showed that bacterial diversity in the sampled mats from Antarctic lakes is extremely high and novel phylotypes were discovered. The observed lower eukaryotic diversity was indeed dominated by a few but highly specialized and often endemic taxa. The overall evidence supported the strategy for industrial screening of these fresh isolates as unexplored source of biotechnologically valuable bioactive molecules.

The main purpose of the studies reported in this paper was to evaluate and compare the ability of Antarctic bacteria and fungi to produce anti-microbial molecules. The frequency of the antimicrobial activities produced by these isolates was considered to be an indicator of their capability to produce anti-infective compounds of potential therapeutic interest. It has to be said that it is always difficult to weight screening results and compare them to those generated by other authors, due to the occurrence of many variable parameters in the microbial product screening process such as the choice of fermentation medium and cultivation conditions as well as the sample preparation methods, the selection of the targets and the conditions of their assay. In this paper we combined screening procedures already in place at Biosearch Italia S.p.A. (now Vicuron Pharmaceuticals) and Merck Research Centers and the resulting frequencies of antimicrobial activities were thus comparable with those routinely achieved by "in house" screening of thousands actinomycetes and fungi isolated from different sources.

Antarctic bacteria have been extensively isolated by using enrichment methods under oligotrophic and psychothropic conditions, from freshwater lakes, saline and hypersaline lakes and ponds, soil, sandstone and sea ice [6, 7, 8, 14, 16, 25, 28, 35, 38, 39, 40, 48]. Several new species have been isolated from Antarctic benthic microbial communities, e.g. *Flavobacterium gelidilacus* [43], *Flavobacterium degerlachei, Flavobacterium frigoris*, *Flavobacterium micromati* [44], *Gillisia limnaea* [45], *Flavobacterium tegenticola* [24], *Arthobacter flavus* [32], *Rhodoferax antarcticus* [22], and the anerobes *Psycrhomonas antarcticus* [27] and *Clostridium vincentii* [26]. Indeed, only few studies were devoted to the high-throughput cultivation and screening of Antarctic microbial isolates [1, 9]. In this work, the 723 Antarctic bacteria belonging to 41 fatty acid clusters and 28 single branches and phylogenetically affiliated with 24 lineages in the alpha, beta and gamma *Proteobacteria*, the *Cytophaga-Flavobacterium-Bacteroides* branch, and the high and low percentage G+C Gram-positives, were massively cultivated and screened for their antimicrobial activities. The nearest validly named phylogenetic neighbours of these strains often belonged to taxa isolated from cold, aquatic environments, such as *Shewanella baltica*, *Psychrobacter glacincola*, *Sulfitobacter pontiacus, Flavobacterium frigidarium, Flavobacterium gillisiae, Saligentibacter salegens, Gelidibacter algens.* Sequences from many strains showed pairwise sequence similarities of less than 97 % to their nearest validly named neighbours [42], indicating that they represent taxa that have not been sequenced yet or as yet unnamed new taxa, related to *Alteromonas*, *Bacillus, Clavibacter*, *Cyclobacterium*, *Flavobacterium*, *Marinobacter*, *Mesorhizobium*, *Microbacterium*, *Pseudomonas*, *Saligentibacter*, *Sphingomonas* and *Sulfitobacter*. These results on the taxonomic diversity and novelty of these isolates supported that they constitute a unique biotechnologically exploitable collection.

From the screening results we observed a high percentage of antibacterial activities (*ca.* 15 %) that contrasts with the few producers of antifungal metabolites (almost 2 %). These frequencies were comparable to the ones observed in the screening of soil actinomycetes, which are considered the most prolific and versatile microbial source of antibiotics [2, 20, 34, 47], supporting the idea that mats from these different lakes contain a rich prokaryotic diversity where the antibiotic production can confer survival advantage. This is indeed not surprising if we consider that these complex microbial communities dominated by prokaryotes have accumulated during thousands of years and bacteria were confronted with extreme conditions, such as low temperatures, freezing-thawing cycles, UV-irradiation, desiccation and varying light conditions, salinities and nutrient concentrations. As a consequence they have been under a high selective pressure and may belong to indigenous new taxa with potentially novel biochemical adaptations.

Distribution of fungi in Antarctica was previously studied mainly in environments such as mosses, lichen communities or in relation to the distribution of "hosts" such as birds, penguins and invertebrates [11, 41, 46]. Most of the filamentous fungi and yeasts described by these authors are cosmopolitan and cold tolerant, but some such as *Thelebolus* appear to be indigenous species [11]. Our investigations [15] revealed that benthic mats are dominated by a relatively small number of fungal species, given the high diversity in eubacteria in the same lakes and compared to the number of species known in the fungal kingdom. None of the filamentous fungi proved to be truly psychrophilic, except *Thelebolus* strains. They often produced markedly pigmented mycelia, probably to protect themselves from strong UV irradiation [18]. Also the production of abundant and resistant spores typical of some of these genera such as *Cladosporium* and *Geomyces* may represent a survival tool to desiccation, to low temperatures or to the presence of high saline concentrations such as it occurs in Vestfold Hills lakes. Finally some species such as *Thelebolus* and *Geomyces* previously isolated by other authors from Antarctic soils or other material such as mosses [11, 41] showed a high tolerance to low temperatures.

The frequency of antibacterial and antifungal activities detected among fungal isolates (29 % active isolates) was higher than the one usually detected in the screening of other ecological groups of fungal isolates [34, 37]. Valuable activities were noted against *C. albicans*, *C. neoformans* or *A. fumigatus* and it is interesting to note that many yeast-like organisms were isolated from Antarctic samples: they often were true psychrophiles and among them were the most frequently isolated genera such as *Candida* and particularly *Cryptococcus* [E. Göttlich and G.S.de Hoog, personal communication; 41]. Possessing metabolic potential to produce active molecules against yeasts can confer a competitive advantage to local filamentous fungi, which turned out to be a further appealing aspect for the biotechnological exploitation of such isolates.

Further work is now in progress on the chemical and biological profiling of metabolites produced by the selected "hits" among these Antarctic fungi and bacteria, which show an interesting antimicrobial spectrum of activities. The assessment of their novelty will help to understand to which extent chemical diversity correlates with the taxonomical diversity so far discovered in these Antarctic benthic microbial communities.

Acknowledgments

This work was supported by the European Community (MICROMAT Project BIO4-98-0040). Thanks for the samplings are due to Kathy Welsh (USA-McMurdo Dry Valleys Long Term Ecosystem Research Program project), Dominic Hodgson and Pip Noon (British Antarctic Survey, UK), Johanna Laybourn-Parry, Gareth Murtagh, Paul Dyer, Ingmar Janse, Tracey Henshaw, Wendy Quayle (University of Nottingham, UK) with logistic support from the Australian Antarctic Division). We thank all the colleagues of MICROMAT consortium and in particular the coordinator Annick Wilmotte, University of Liège, Belgium.

Authors declare that the experiments comply with the current laws of the country in which they were performed.

References

1. Ashbolt, N.J. (1990). Antarctic biotechnology-what is the potential? *Australian Journal of Biotechnology*, **4**:103-105.
2. Axelrood, P.E., Clarke, A.M., Radley, R., and Zemcov, J.V. (1996). Douglas-fir root-associated microorganisms with inhibitory activity towards fungal plant pathogens and human bacterial pathogens. *Canadian Journal of Microbiology,* **42**: 690-700.
3. Basilio, A., Pelaez F., Diez, M.T., Vicente, F., Morris, S.A., Curotto, J.E., Bills, G.F., Dreinkorn, S.J., Hensens, O.D., Liesch, J.M. and Thompson, J.R. (1998). *Antifungal agents*. U.S. patent 5,770,587.
4. Bernan, V.S., Greenstein, M. and Maiese, W.M. (1997). Marine microorganisms as a source of new natural products. *Advanced Applied Microbiology,* **43**:57-90.
5. Bills, G.F., Diez, M.T., Omstead, M.N. and Pelaez, F. (1993). *Processes for preparing novel squalene synthetase inhibitors*. U.S. patent 5,250,424.
6. Bowman, J.P., McCammon, S.A., Brown, M.V., Nichols, D.S., and Mcmeekin, T.A. (1997b). Diversity and association of psychrophilic bacteria in Antarctic sea ice. *Applied Environmental Microbiology,* **63**:3068-3078.
7. Bowman, J. P., Rea, S. M., and McMeekin, T. A. (2000). Diversity and community structure within anoxic sediment from marine salinity meromictic lakes and a coastal meromictic marine basin, Vestfold Hills, Eastern Antarctica. *Environmental Microbiology,* **2**:227-237.
8. Brambilla, E., Hippe, H., Hagelstein, A., Tindall, B.J. and Stackebrandt, E. (2001). 16S rDNA diversity of cultured and uncultured prokaryotes of a mat sample from Lake Fryxell, Mc Murdo Dry Valleys, Antarctica. *Extremophiles,* **5**:23-33.
9. Bull, A.T., Ward, A.C. and Goodfellow, M. (2000). Search and discovery strategies for biotechnology: the paradigm shift. *Microbiology and Molecular Biology Reviews,* **64**:573-606.

10. Carelli, A., Borghi, A., Carrano, L., Gastaldo, L. and Marinelli, F. (1995). Biotransformation of lipoglycopeptide antibiotic A40926 by *Actinoplanes teichomyceticus* cells. *Journal of Industrial Microbiology,* **15**:429-433.

11. Del Frate, G., and Caretta, G. (1990). Fungi isolated from antarctic material. *Polar Biology,* **11**:1-7.

12. Dombrowski, A., Jenkins, R., Raghoobar, S., Bills, G., Polishook, J., Pelaez, F., Burguess, B., Shao, A., Huang, L.Y., Zhang, Y. and Goetz, M. (1999). Production of a family of kinase-inhibiting lactones from fungal fermentations. *Journal of Antibiotics* (Tokyo), **52**:1077-1085.

13. Doran, P.T., Wharton, R.A. Jr. and Lyons, W.B. (1994). Paleolimnology of the McMurdo Dry Valleys, Antarctica. *Journal of Paleolimnology,* **10**:85-114.

14. Friedmann, E.I. (Ed.) (1993). *Antarctic Microbiology.* Wiley-Liss Inc, New York.

15. Göttlich, E., de Hoog, G. S., Genilloud, O., Jones, B. E. and Marinelli, F., (2003). MICROMAT: Cultivable diversity in fungi found in microbial mats of Antarctic lakes. *Book of Abstracts Antarctic Biology in a Global Context, VIII SCAR International Biology Symposium*, Amsterdam, The Netherlands, 27 August-1 September 2001, n°S5P21.

16. Gonsink, J.J. and Staley, J.T. (1995). Biodiversity of gas vacuolated bacteria from Antarctica sea ice and water. *Applied Environmental Microbiology,* **61**:3486-3489.

17. Hall, R.M., Dawson, M.J., Jones, C.A., Roberts, A.D., Sidebottom, P.J., Stead, P. and Taylor, N.L. (2001). The production of novel sordarin analogues by biotransformation. *Journal of Antibiotics* (Tokyo), **54**:948-957.

18. Hughes, K.A., Lawley, B. and Newsham, K. (2003). Solar UV-B radiation inhibits the growth of Antarctic terrestrial fungi. *Applied Environmental Microbiology,* **69**:1488-1491.

19. Laybourn-Parry, J., James, M.R., McKnight, D., Priscu, J., Spaulding, S. and Shiel, R. (1997). The microbial plankton of Lake Fryxell, Southern Victoria land, Antarctica. *Polar Biology,* **17**:54-61.

20. Lazzarini, A., Cavaletti, L., Toppo, G. and Marinelli, F. (2000). Rare genera of actinomycetes as potential producers of new antibiotics. *Antonie van Leeuwenhoek,* **78**:399-405.

21. MacNeil, I.A., Tiong, C.L., Minor, C., August, P.R., Grossman, T.H., Loiacono, K.A., Lynch, B.A., Phillips, T., Narula, S., Sundaramoorthi, R., Tyler, A., Aldredge, T., Long, H., Gilman, M., Holt, D. and Osburne, M.S. (2001). Expression and isolation of antimicrobial small molecules from soil DNA libraries. *Journal of Molecular Microbiology and Biotechnology,* **3**: 301-308.

22. Madigan, M.T., Jung, D.O., Woese, C.R. and Achenbach, L.A. (2000). *Rhodoferax antarcticus* sp.nov., a modaretely psychrophilic purple nonsulfur bacterium isolated from an Antarctic microbial mat. *Archives of Microbiology,* **173**:269-77.

23. Marinelli, F., Stella, S., Montanini, N., Le Monnier, F., Cavaletti, L. and Selva, E., (1995). Bioconversion capacity of *Streptomyces* sp. GE44282, a producer of the antibiotics heneicomycin and aurodox. *Journal of Industrial Microbiology,* **15**:49-54

24. McCammon, S.A. and Bowman, J.P. (2000). Taxonomy of Antarctic *Flavobacterium* species: description of *Flavobacterium gilissiae* sp.nov., *Flavobacterium tegetincola* sp.nov. and *Flavobacterium xanthum* sp.nov., nom.rev., and reclassification of [*Flavobacterium*] *salengens* as *Saligentibacter salegens* sp.nov., comb.nov. *International Journal of Systematic Bacteriology,* **50**:1055-1063.

25. Mergaert, J., Verhelst, A., Cnockaert, M.C., Tan, T.L. and Swings, J. (2001). Characterization of facultative oligotrophic bacteria from polar seas by analysis of their fatty acids and 16S rDNA sequences. *Systematic and Applied Microbiology,* **24**: 98-107.

26. Mountfort, D.O, Rainey, F.A., Burghardt, J., Kaspar, H.F. and Stackebrandt, E. (1997). *Clostridium vincentii* sp.nov, a new obligately anaerobic halophilic psychrophile isolated from pond sediment of the McMurdo Ice Shelf, Antarctica. *Archives of Microbiology,* **169**: 231-238.

27. Mountfort, D.O., Rainey, F.A., Burghardt, J., Kaspar, H.F. and Stackebrandt, E. (1998). *Psychromonas antarcticus* sp.nov., a new aerotolerant anaerobic saccharolytic psychrophilic bacterium isolated from low salinity pond sediment of the McMurdo Ice Shelf, Antarctica. *Archives of Microbiology,* **167**: 54-60.

28. Murray, A.E. (1998). Seasonal and spatial variability of bacterial and archaeal assemblages in the coastal waters near Anvers Island, Antarctica. *Applied Environmental Microbiology,* **64**:2585-2595.

29. Narinx, E., Baise, E. and Gerday, C. (1997). Subtilisin from psychrophilic Antarctic bacteria: characterization and site-directed mutagenesis of residues possibly involved in the adaptation to cold. *Protein Engineering,* **10**:1271-1279.

30. Obata, H., Muryoi, N., Kawahara, H., Yamade, K. and Nishikawa, J. (1999). Identification of a novel ice-nucleating bacterium of Antarctic origin and its ice nucleation properties. *Cryobiology,* **38**:131-139.

31. Pelaez, F. and O. Genilloud. (2002). Discovering new drugs from microbial natural products. In *Microorganisms for health care, food and enzyme production*, (Barredo, J.L., Ed.), Research Signpost.

32. Reddy, G.S., Aggarwal, R.K., Matsumoto, G.I. and Shivaji, S. (2000). *Arthobacter flavus* sp.nov, a psychrophilic bacterium isolated from a pond sediment in McMurdo Dry Valleys, Antarctica. *International Journal of Systematic Bacteriology,* **50**: 1553-1561.

33. Rondon, M. R., Goodman, R. M. and Handelsman, J. (1999). The Earth's bounty: assessing and accessing soil microbial diversity. *Trends in Biotechnology,* **17**: 403-409.

34. Sponga, F., Cavaletti, L., Lazzarini, A., Losi, D., Borghi, A., Ciciliato, I. and Marinelli, F. (1999). Biodiversity and potentials of marine-derived microorganisms. *Journal of Biotechnology,* **70**:65-69.

35. Staley, J.T. and J.J. Gosink (1999) Poles apart: biodiversity and biogeography of sea ice bacteria. *Annual Review of Microbiology,* **53**:189-215.

36. Stefanelli, S,, Sponga, F., Ferrari, P., Sottani, C., Corti, E., Brunati, C., and Islam, K. (1996). Inhibitors of myo-inositol monophosphatase, ATCC 20928 factors A and C. Isolation, physico-chemical characterization and biological properties. *Journal of Antibiotics* (Tokyo), **49**:611-616.

37. Suay, I., Arenal, F., Asensio, F.J., Basilio, A., Cabello, M.A., Díez, M.T., García, J.B., González del Val, A., Gorrochategui, J., Hernández, P., Peláez, F. and Vicente, M.F. (2000). Screening of basidiomycetes for antimicrobial activities. *Antonie van Leeuwenhoek,* **78**:129-139.

38. Tan, T.L., Reinke, M. and Rüger, H.J. (1996). New diluition method in microtiter-plates for enumeration and enrichment of copiotrophic and oligotrophic bacteria. *Archiv fur Hydrobiologie,* **137**:511-521.

39. Tan, T.L., Reinke, M. and Rüger, H.J. (1999). Enrichment, isolation, and Biolog metabolic fingerprints of oligotrophic bacteria from the Antarctic Ocean. *Archiv fur Hydrobiologie,* Special Issue. *Advances in Limnology,* **54**:255-272.

40. Tindall, B., Brambilla, E., Steffen, M., Neumann, R., Pukall, R., Kroppenstedt, R.M. and Stackebrandt, E. (2000). Cultivable microbial biodiversity: gnawing at the Gordian knot. *Environmental Microbiology,* **2**: 310-318

41. Tosi, S., Casado, B., Gerdol, R. and Caretta, G. (2002). Fungi isolated from Antarctic mosses. *Polar Biology,* **25**: 262-268

42. Van Trappen, S., Mergaert, J., Van Eygen, S., Dawyndt, P., Cnockaert, M.C. and Swings, J. (2002). Diversity of 746 heterotrophic bacteria isolated from microbial mats in Antarctic lakes. *Systematic and Applied Microbiology,* **25**: 603-610.

43. Van Trappen, S., Mergaert, J. and Swings J. (2003). *Flavobacterium gelidilacus* sp. nov. isolated from microbial mats in Antarctic lakes. *International Journal of Systematic and Evolutionary Microbiology,* **53**: 1241-1245.

44. Van Trappen, S., Vandecandelaere, I., Mergaert, J. and Swings, J. (2004). *Flavobacterium degerlachei* sp. nov., *Flavobacterium frigoris* sp. nov. and *Flavobacterium micromati* sp. nov., novel psychrophilic bacteria isolated from microbial mats in Antarctic lakes. *International Journal of Systematic and Evolutionary Microbiology,* **54**: 85-92

45. Van Trappen, S., Vandecandelaere, I., Mergaert, J. and Swings, J. (2004). *Gillisia limnaea* gen. nov., sp. nov., a new member of the family *Flavobacteriaceae* isolated from a microbial mat in Lake Fryxell, Antarctica. *International Journal of Systematic and Evolutionary Microbiology,* **54**: 445-448

46. Vishniac, H.S. (1993). The microbiology of Antarctica soils. In *Antactic Microbiology* (Friedmann, E.I., Ed.). Wiley, New York. pp.297-341.

47. Waksman S. A. and Lechevalier, H. A. (1962). The search for antibiotics: screening programs. In *The Actinomycetes,* Vol.3 Antibiotic of actinomycetes (S. A. Waksman, and H. A. Lechevalier, Eds.). Williams & Wilkins Company, Baltimore, pp.20-39.

48. Wery, N., Gerike, U., Sharman, A., Chaudhuri, J.B., Hough, D.W. and Danson, M. (2003). Use of a packed-column bioreactor for isolation of diverse protease-producing bacteria from Antarctic soil. *Applied and Environmental Microbiology,* **69**: 1457-1464.

49. Wharton, R.A. Jr., Lyons, W.B. and Des Marais, D.J. (1993). Stable isotopic biogeochemistry of carbon and nitrogen in a perennially ice-covered Antarctic lake. *Chemical Geology,* **107**: 159-172.

Marine microbes for biodiscovery: Just the tip of an iceberg

D. G. Bourne, E. Evans-Illidge and L. E. Llewellyn

Australian Institute of Marine Science, PMB 3, Townsville, Queensland Australia, 4810

1. Introduction

There are many definitions for biodiversity, though at its simplest, it refers to a catalogue of living species within a certain habitat or ecosystem. Although this term is used extensively in microbiology, understanding microbial biodiversity in even the simplest habitats is enormously difficult. For an environment as large and diverse as the marine ecosystem, an understanding of microbial biodiversity is therefore complex and challenging. Some estimate the entire microbial diversity of the sea at around 2 x 10^6 species [17]. This is an enormous reservoir of organisms, many of which possess unique biological facets that represent an unexplored reservoir of natural chemicals and other molecules awaiting discovery.

Prokaryotes represent a large proportion of the earth's biota, though this great biodiversity remains virtually untapped. Presently, only about 5000 microorganism species are reported as cultured [15, 32] and there is little doubt our knowledge of the real extent of prokaryotic diversity is indeed small. A large part, if not most, is hidden in nearly every conceivable environment, unable to be grown in pure culture. The majority of our knowledge of the uncultured biodiversity derives from studies of 16S rDNA sequences. To date, 53 bacterial phyla have been identified, though only 27 of these phyla contain representatives that have been cultivated [32, 42]. Of these, many phyla contain only a few representative grown strains and some contain only 1 described species. Five phyla, *Actinobacteria*, *Bacteroidetes*, *Cyanobacteria*, *Firmicutes* and *Proteobacteria* contain 95% of all cultivated and published bacterial species [32].

The marine environment is home to representatives of most of these known taxonomic phyla, with organisms inhabiting a realm of extremes of temperature, pressure, light and salinity beyond tolerable boundaries of other life forms. The physiological responses of populations to cope with these extreme environments lead to novel metabolite chemistry arising from metabolic processes unmatched in the laboratory, and which can produce chemical structures unimagined by scientists. This potential novel marine metabolic chemistry is a key to the re-emerging interest in natural products chemistry as a source of antibiotics, drugs, herbicides, insecticides and fungicides to combat many of the world's diseases and afflictions. Adding to this new wave of investigations in marine microbial biodiversity and novel microbial metabolic chemistry is the new culture-based and molecular tools now available to investigate and harness microbial biodiversity.

The potential harvest of natural products, pharmaceutical drugs, and other chemical compounds derived from the marine environment has yet to be realised. Almost 50% of modern drugs are derived from natural products [39] though virtually all these are from a terrestrial origin [33]. However in recent years, marine natural product biodiscovery has yielded a considerable number of drug candidates, most still in pre-clinical or early clinical development (Table 1) [26]. Some marine natural products have reached the market, such as cytarabine, derived from C-nucleosides from the Caribbean sponge, *Cryptotheca crypta* [44], a chemotherapeutic drug used to treat acute myeloid leukaemia [41]. Cephalosporin antibiotics were also first derived from pseudo-marine fungi [40]. Other commercialised natural products from the ocean include the nereistoxin analogue insecticides, cartap, bensaltap and thiocyclam, developed from the Japanese marine worm *Lumbriconereis heteropoda* [36]. Green fluorescent protein (GFP), cloned from the jellyfish *Aequorea victoria,* is another famous example of a marine natural product used commercially for a wide range of applications. These include reporter systems for environmental microbiology research [21], plant genetic transformation studies [47] and insect transgenesis [28] to name a few.

2. The importance and scope of marine microbes

The role microbes play in the marine environment is complex and diverse depending on the ecological niche sustaining the organisms. Within the open oceans, microbial cells are divided among three domains of life: Eukaryota, Archaea and Bacteria. Small eukaryotic organisms ($<3\ \mu m$), make up a large proportion of the oceanic picoplankton photosynthesising in the open ocean [45]. Pelagic archaea comprise a large proportion of the total marine picoplankton, being as much as 40% of detected microbial diversity at certain depths [24, 31, 37]. The final group, Bacteria, performs a wide range of metabolically diverse activities, most of which we are just now starting to understand. For example, assemblages of widely distributed, distantly related, photosynthetically active bacterial groups, including some unrelated to known and cultivated types, have recently been identified in marine plankton and carry out aerobic, anoxygenic, photosynthesis [5]. This photosynthetic pathway was only discovered 20 years ago [46] and now may account for 11% of the total microbial community in the oceans [5]. These microbial populations within the oceans and their metabolic processes ultimately influence the earth's climate by driving carbon and nitrogen cycling in the oceans. A fourth group of biological entities, the viruses are not organisms in the same sense as the Eukaryota, Archaea and Bacteria. However, they are of considerable biological importance, being the most abundant biological entities in the sea (10^7 ml^{-1}) [2]. Oceanic bacterial numbers and diversity are kept under control by bacterial specific viruses (bacteriophages), a process central to upper ocean respiration through the constant cycling of carbon as a result of bacterial assimilation of dissolved organic matter and subsequent release by bacterial lysis and metabolism by other bacterial populations [23].

Open oceans are just one environment however within the marine ecosystem, with marine sediments and subsurface communities, deep-sea hydrothermal vents and mangroves providing an array of niches for microbial communities. This list does not include symbiotic associations between microbes and macro-organisms, with the biomass present in many marine animals being far more complex than may appear on the surface. Microscopic symbionts are critical to the fundamental physiology, if not survival, of their marine host. The most well known example is corals, which have a mutualistic relationship with zooxanthellae. The coral provides the algae with a protected environment and compounds they need for photosynthesis while in return, the algae produce oxygen and help the coral to remove wastes. Most importantly, zooxanthellae supply the coral with glucose, glycerol, and amino acids, which are the products of photosynthesis. The coral uses these products to make proteins, fats, and carbohydrates, and produce calcium carbonate [3, 4, 34, 35]. The relationship between the algae and coral polyp facilitates a tight recycling of nutrients in nutrient-poor tropical waters [48].

Another example is the interaction of sponges and bacteria, many of which produce a suite of potent and novel chemical compounds. This host/symbiont association can make invertebrates the biofactories for bioactive metabolites manufactured by the symbiont, or the invertebrates may metabolically modify the microbial natural product, or *vice versa*. In fact, the existence of the symbiosis may be critical to bioactivity production with neither the host nor the symbiont able to manufacture the chemicals in isolation. Therefore there is enormous potential for new active metabolites and novel chemical pathways within marine microorganisms and their hosts. Many of the bioactive molecules however, are believed to be derived from the symbiont microorganism. The best-known example of this is the protein kinase C inhibitors called bryostatins currently in clinical trials as anti-cancer drugs (Table 1). These chemicals were derived from the bryozoan, *Bugula neritina*, though it has long been suspected an uncultivated symbiotic γ-proteobacterium produces the molecules as secondary metabolites [18].

Although the potential is large, limitations still exist for harnessing the microbial biodiversity due to the inability to culture the organism. Ultimately a comprehensive understanding of the physiology of organisms can only be obtained by its cultivation. There is an important bonus in the discovery of commercial products from culturable microorganisms. If you can culture them, you also have the means of economic, sustainable, large-scale production. In terms of total microbial diversity, typically less than 1% of the bacteria present in the marine environment can be cultivated in the laboratory. Yet we have learned much about microbial processes by studying less then 99% of these microorganisms. No doubt there is a cache of important new principles, metabolic pathways and novel chemicals lurking in the 99% of the microbial world that is yet untapped.

3. Australian Institute of Marine Science (AIMS) Biodiscovery Project

Australia is particularly biologically rich being one of 12 recognised mega-biodiverse nations, however only a small fraction of this diversity has been explored and microbial diversity has remained almost untouched. The diversity in some regions is highly endemic, having evolved in the 200 million years since the separation of Australia from prehistoric Gondwanaland and its drift into the disappearing Tethys Sea. Species diversity equates to chemical diversity and builds on 4 billion years of evolution. This diversity is magnified as metabolite chemistry alters as a result of physiological responses to variation in exogenous factors associated with abiotic and biotic changes in habitat, season and location.

The Australian Institute of Marine Science is a statutory authority of the federal government of Australia headquartered in Townsville (Figure 1) to be central to one of the world's largest and most biodiverse ecosystems, the Great Barrier Reef. AIMS also has a research base in Western Australia (Figure 1), and are expanding its efforts across northern Australia with a new laboratory in Darwin (Figure 1).

Table 1: Marine natural products and derivatives in clinical development as pharmaceuticals

Compound	Source	Application	Status at March, 2004
Bryostatin 1	*Bugula neritina* (bryozoan)	Anti-tumour	Phase I & II as a co-drug with other anticancer drugs such as paclitaxel, vincristine, cisplatin and interleukin-2
Dolastatin 10	*Dolabella auricularia* (mollusc)	Anti-tumour	Phase II clinical trials in USA
Ecteinascidin 743	*Ecteinascidia turbinate* (ascidian)	Anti-tumour	Phase II trials in USA Phase III complete in Europe but not authorised for marketing
Aplidine	*Aplidium albicans* (ascidian)	Anti-tumour	Phase II clinical in Europe and USA
Kahalalide F	*Elysia rubefescens* (mollusc)	Anti-tumour	Phase II clinical trials in Europe
KRN-7000	*Agelas mauritianus* (sponge)	Anti-tumour	Phase I clinical trials in USA
Methopterosin #	*Pseudopterogorgia elisabethae* (soft coral)	Anti-inflammatory	Phase I clinical trials
IPL-576092 #	*Petrosia contignata* (sponge)	Anti-asthmatic	Phase I clinical trials
Manoalide	*Luffariella variabilis* (sponge)	Anti-inflammatory	Phase I clinical trials
Squalamine lactate	*Squalus acanthias* (shark)	Anti-tumour	Phase I clinical trial in USA when administered alone Phase II clinical trials in USA when administered with carboplatin
GTS-21 #	*Amphiporus lactifloreus* (worm)	Alzheimer's Disease	Phase I clinical trials
Prialt (= ziconitide)	*Conus magus* (mollusc)	Chronic pain	Phase III complete in USA, subject of New Drug Application Drug awaiting Marketing Authorization in Europe

synthetic analogue of marine natural product lead

Underpinning its biotechnology are strengths in:

- ☑ taxonomy,
- ☑ marine ecology,
- ☑ database design and data mining,
- ☑ bioassay and biodetection
- ☑ aquaculture of marine animals in aquaria, pools, dams and the ocean
- ☑ cell culture of both microbes and tissues
- ☑ gene discovery
- ☑ expression of recombinant proteins
- ☑ biomolecule purification and identification

Figure 1: Australian coastline depicting location of major coral reef complexes and the location of the three laboratory complexes of the Australian Institute of Marine Science, Townsville, Darwin and Perth

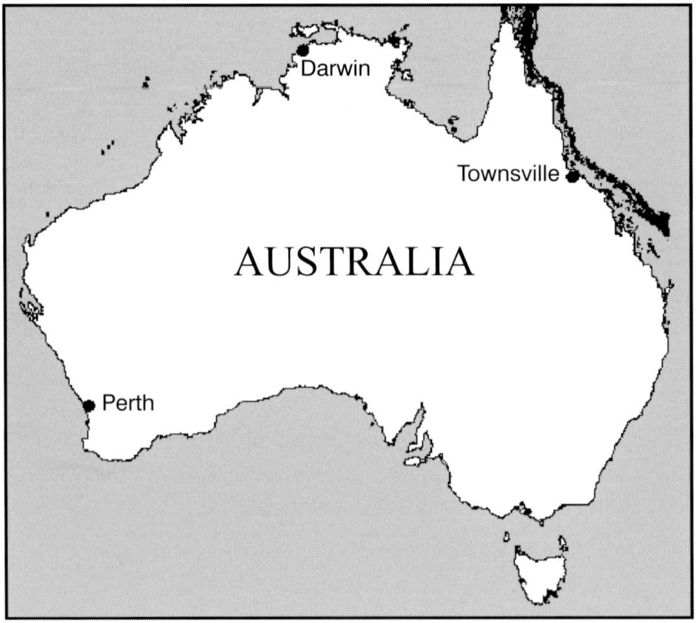

AIMS has a biodiscovery collection which includes material from around 10,000 marine macroscopic organisms (eg corals, sponges, sea squirts) and 9,000 marine microorganisms, collected and isolated from over 1500 sites from around Australia (Figure 2). The macroorganism collection includes all major marine phyla, including Porifera (36% of library), Cnidaria (16%) and Chordata (11%) (Figure 3). The collection represents a wide biodiversity in terms of taxonomy, ecology, and geography and was designed for natural products research, but also includes material and data for systematic and taxonomic study of the samples. It is associated with a comprehensive relational database, and material includes frozen biomass, taxonomic vouchers and a library of extracts for screening.

Figure 2: Geographical location map of sites sampled around Australia for biodiversity and biodiscovery collection at AIMS

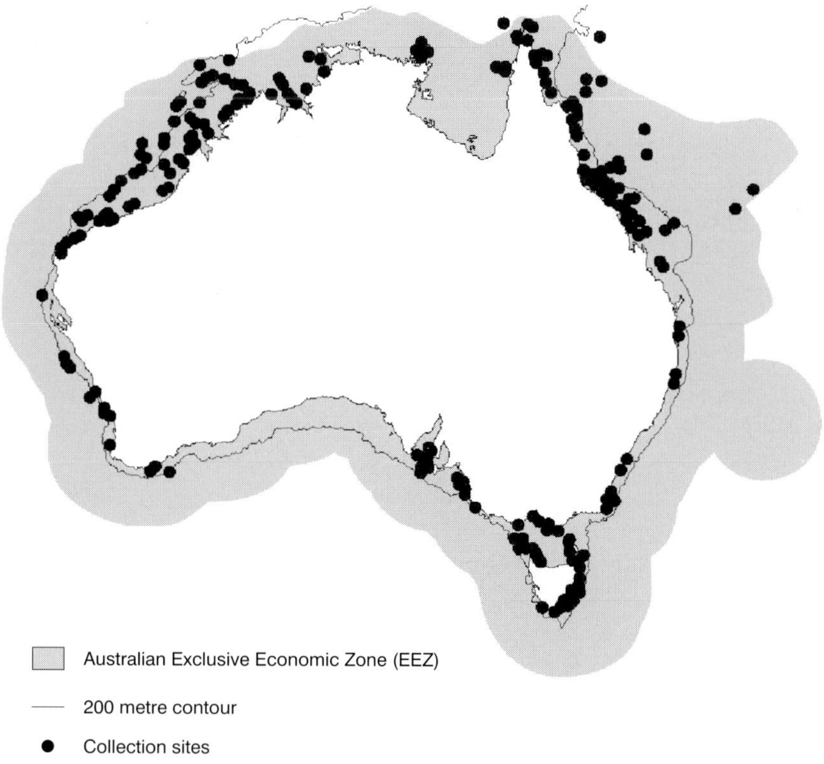

Figure 3: Taxonomic breakdown of the AIMS macro-organism biodiversity collection

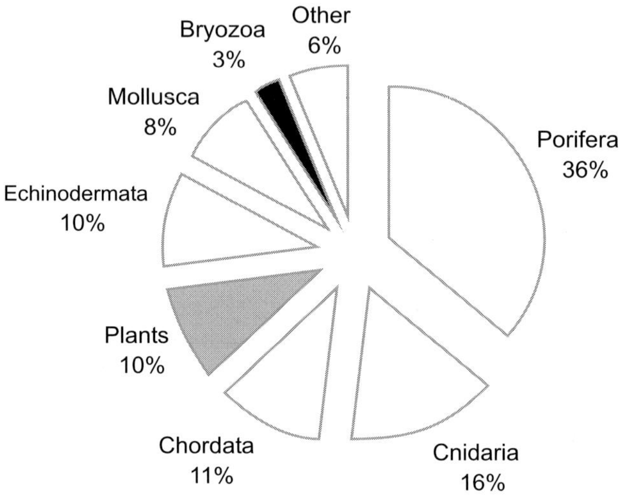

Figure 4: Taxonomic breakdown of the AIMS microorganism biodiversity collection

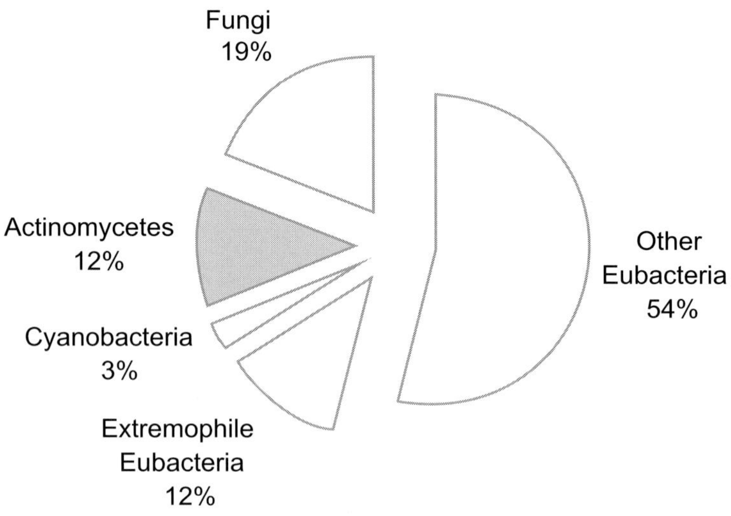

kingdom Eubacteria (66% of library, including extremophile isolates) and complemented by Actinomycetes (12%) and Cyanobacteria (3%) within this kingdom (Figure 4). The other major group of organisms present in the collection is Fungi (19% of library). These organisms have been isolated from a variety of environments using selective and general classical microbiological media. Organisms are isolated by taking specific swabs of samples and direct plating, from the water column, various sediments and unique niches on the surfaces of marine organisms such as coral mucous. As highlighted previously, many natural products from marine invertebrates are, or are suspected to be, of microbial origin [18]. As a result, tremendous effort has been invested in the isolation of several thousand microbial isolates from the tissues of a subset of marine invertebrates, which may be either symbiotic or opportunistic occupants of the organism. This is achieved by crushing whole samples before plating onto microbiological media. A wide range of media types have been used for isolation of organisms, some being general media formulations such as Marine Agar for isolation of a broad range of heterotrophic marine bacteria, others being specific for groups of organisms such as TCBS for the *Vibrionaceae* and *Alteromonas* groups of organisms. Potato Dextrose Agar has been used for cultivation of mould and yeasts while Malt Extract Agar and appropriate antibiotics can be used for selective isolation and cultivation of fungi. Microbial isolates undergo successive sub-culturing to ensure that single colonies representative of pure strains are obtained. Once the isolates appear as pure strains they are assigned phenotypic characteristics including descriptions of colony colour, optical appearance, colony form/shape, colony margin, colony elevation, colony surface, colony texture and colony size. All materials are cryopreserved and ready for molecular, taxonomic, and metabolite analysis, as well as novel enzyme discovery.

A subset of these microbes have already been used to develop a library of ~5,000 extracts for new chemical screening by fermenting in multiple liquid media which expose the microbe to conditions such as high and low pH, high and low salt, and high and low nutrients. These mini-fermentations are then extracted and screened for biological activity.

All descriptions and information regarding collection, isolation and screening for both macro and microorganisms are entered into a comprehensive relational database. Figure 5 provides an example of how this database operates, with searchable interfaces at all levels. Microbial housekeeping information includes, phylogenetic or taxonomic information, isolation procedures, storage procedures and colony morphology information. Each record is linked to an image to confirm that an isolate recovered from cryopreservation matches the original isolate. Each record also links to other elements of the biodiscovery process. Collection site details such a longitude and latitude measured by GPS as well as descriptions of the site are recorded to enable re-collection. If the organism was isolated from a macroscopic organism, all information related to the collection, location and isolation procedures are stored. Information relating to fermentation of the microorganism for active metabolite extraction is archived in the database, including fermentation media used, temperatures, pH and extraction protocol. Finally, bioassay results and all chemistry related information is linked to the isolate.

3.1 AIMS Screening Targets

A range of targets have been chosen for screening of the microbial metabolite extract library, including tumour cell cytotoxicity (in collaboration with National Cancer Institute in the United States), herbicide screening both at the enzymatic assay level and whole plant assays plus anti-infective screening. Herbicide screening using whole plant assays and anti-infective screening will be discussed in more detail subsequently.

3.1.1 Whole plant phytotoxicity assays

In contrast to the search for new pharmaceuticals, little research effort has explored marine natural products for agrochemical compounds [22]. However marine areas and particularly coral reefs appear to be an ideal environment to search for herbicide candidates since animals (*i.e.* corals and other reef building organisms) are in a constant chemical warfare with plants (*i.e.* algae) for available space and nutrients [36]. As part of a proof of concept research, AIMS has investigated the phytotoxicity of marine microbial extracts applied to millet and oat seedlings. Plants can be divided into three main groups depending on how they fix inorganic carbon into sugars, C_3-, C_4- and crassulacean acid metabolism (CAM) plants [1]. The world's most important crops are C_3 plants (rice, wheat, oats and barley). However 14 of the world's top 18 weeds are C_4 plants [27]. Therefore using Oat plants (C_3 photosynthesis pathway) and millet (C_4 pathway) as model systems, whole plant assays are performed to identify microbial extracts with activity to these plants. This low technology assay provides an effective means for active extract identification that can subsequently be analysed by high technology chemistry to identify active compounds that specifically inhibit photosynthetic pathways of the plants. While slower and requiring larger amounts of test material than enzymatic assays, the whole plant assays are information rich and identify those bioactive chemicals that can penetrate the plant's barriers such as the cuticle and cell wall.

3.1.2 Anti-infective screening strategy

The urgent need for the discovery of new antimicrobials has been highlighted as one of the needs of humankind in the 21st century [7]. While antibiotic resistance in bacteria increases, the pace at which new antibiotics are being produced is slowing. Reports of methicillin-resistant *Staphylococcus aureus* (MRSA), vancomycin-resistant enterococci (VRE), multi-drug resistance in *Mycobacterium tuberculosis* (MDRTB) strains and multi-drug-resistant (MDR) Gram-negative bacteria are commonplace [13, 38]. New and emerging pathogens add to the gravity of the situation. Apart from antibiotics taken as human drugs, antimicrobial resistance is selected for by the use of antibiotics as crop protection agents and in treating food and companion animals which, if not chemically identical, act on the same mechanisms as those used to treat humans. Diversification of the antibiotics arsenal is an agreed strategy to reduce the emergence of antibiotic resistance allowing new antibiotics to remain effective for longer periods. This need is reflected in the high priority given to anti-infective discovery by the major pharmaceutical companies in the world.

As a precursor to screening against pathogenic or model microbes, AIMS has screened its extract collection against the "lab-rats" of the microbiological world, namely *Escherichia coli*, *Staphylococcus aureus* and *Candida albicans*. This dataset (ie activity profile against *E. coli*, *S. aureus* and *C. albicans*) while valuable in its own right, becomes immensely valuable when used to ground-truth screening against new anti-infective targets. A critical gap in antibiotic discovery is to identify chemicals that may act via new mechanisms of action [52].

Extracts which lack activity against these three lab-rat microbes yet potent activity against the anti-infective target of choice will provide a significant early indication that the "hit" organism may contain a molecule or molecules that acts via a mechanism not present in *E. coli*, *S. aureus* and *C. albicans* (Table 2).

This novelty would enable prioritisation of the investigative effort upon these unique bioactive leads. In some cases, this can be coupled with one of the genomics efforts elsewhere in the world to allow data mining and bioinformatics approaches to identify the possible molecular site of action. For example, the fast growing and relatively benign mycobacterium, *Mycobacterium smegmatis*, is the subject of a

Figure 5: Relational database for storage of all information pertaining to the collection, maintenance and exploitation of the AIMS biodiversity collection

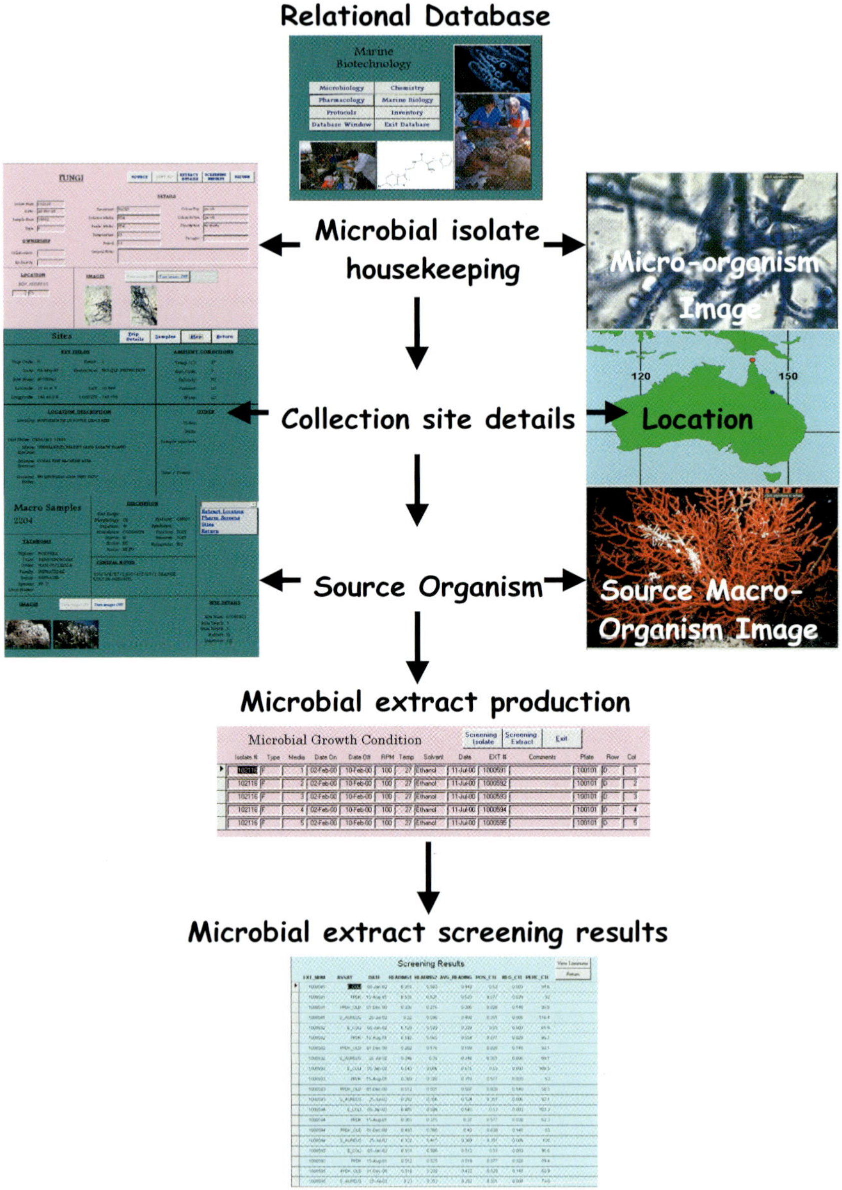

genome effort at the Institute for Genomic Research in Maryland (www.tigr.org). This microbe is used as a proxy for other pathogenic mycobacteria such as *M. leprae* (the cause of leprosy), *M. tuberculosis* (the causative agent of tuberculosis) and *M. ulcerans* (the cause of Buruli ulcers), all significant health problems of the developing world.

Table 2: Selective criteria for prioritisation of anti-infective screening extracts

Escherichia coli	***Staphylococcus aureus***	***Candida albicans***	**New target**	**Interpretation**
Active	Active	Active	Active	General toxicity likely
Inactive	Inactive	Inactive	Active	Potential novel anti-infective lead acting through a novel mechanism

4. Other application of marine microbes

4.1 Bio-indicators

Exploitation and commercial use of marine microorganisms is not limited to pharmaceutical and agrochemical discovery. New tools are becoming available to investigate not only microbial community structure but also microbial function within that community. It is now possible to link microbial diversity with activity and establish the ecological function these organisms perform within their environment. This information can be extrapolated to determine the roles these organism have in whole ecosystems and their impact on global cycles. An understanding of this functional role of microorganisms within marine ecosystems is vital for understanding how ecosystems sustain themselves and allows better management of environmental problems that may impact on that ecosystem, such as global warming.

It is becoming increasingly apparent that microbial activity and function is fundamental to ecosystem health [2]. Changes in microbial community structure may provide the first indication of perturbations and/or stress on an ecosystem. Therefore microbial populations may act as sentinels for environmental health and to some extent, microbial populations have played this role for a long time. Cyanobacterial blooms in rivers or red-tide blooms in the oceans are indicators of an unbalanced ecosystem possibly through excessive nutrient inputs or other environmental impacts. With the advent of new technologies such as microarrays, we can follow microbial populations and microbial genes within an environment and potentially link microbial community structure and function to environmental conditions. Although in its infancy in technology development, exploiting microbial changes has enormous potential for producing environmental sensors and bio-indicators reflecting environmental impacts.

One example where such an approach has enormous potential is for investigating coral health and identifying coral diseases. Rohwer *et al.* [43] identified 430 novel bacterial ribotypes associated with three coral species. These bacterial species appear to have symbiotic relationships with the animal, and corals may in fact harbour specific microbial communities for beneficial effect. Shifts in the bacterial species composition affect the coral health and susceptibility to disease. Therefore, once the normal microbial flora of corals and pathogen(s) causing a specific coral disease are identified, it is then potentially possible to assess changes in microbial populations in complex environments like coral tissue. These changes in the microbial community can be monitored and the appearance of certain microbial species may provide an early warning of potential coral disease outbreak. A similar project at AIMS is investigating microbial communities and response of these communities as bio-indicators of nutrient input into reef systems.

5. Novel Methodologies for Biodiscovery

5.1 Culture-based

Traditional culture-based isolation strategies utilise an environment very different to that which most microorganisms occupy within nature and are therefore biased towards certain groups of the microbial world. This explains why 95% of all cultivated and published bacterial species fall within the five bacterial phyla, *Actinobacteria*, *Bacteroidetes*, *Cyanobacteria*, *Firmicutes* and *Proteobacteria* [42]. These organisms tend to be fast-growing, high density isolates resistant to high nutrient concentrations and able to grow in isolation. In contrast, most marine bacteria exist in an environment that is oligotrophic, and react differently to cultural-confinement, substrate quality and quantity, nutrient availability and presence of toxin products and viral infections. Most cells that are observed in the marine environment are viable however, and therefore potentially culturable [20]. To isolate these organisms, an appropriate cultivation strategy is required which depends on an understanding of the physiological requirements of the organism. In an attempt to increase our known culturable microbial diversity, a range of new and exciting cultivation methods have been, and are, being developed [8, 12, 14, 20, 30, 54].

Many of the new cultivation techniques attempt to simulate the natural environment by diluting growth media to support low levels of cell growth and allow long incubation times. For example, Cho and Giovannoni [12], isolated forty-four novel strains of *Gammaproteobacteria* from pelagic and oceanic Pacific Ocean regions using high-throughput culturing methods. This approach relies on bacterial dilution to extinction in very low nutrient media, which represents the chemical composition of the microbes' natural marine environment. Typically, the substrate concentration is 3 orders of magnitude less than commonly used media and has been used to isolate many microbial strains including members of previously uncultured groups believed to be abundant in coastal waters [14]. Other studies have placed microorganisms in diffusion chambers and incubated the chambers in an aquarium that simulated these organisms natural environment [30]. The membranes of the sealed chambers allow exchange of chemicals between the chamber and the environment but restrict movement of cells. Another method encapsulates single cells in gel microdroplets that are then cultivated under low nutrient flux conditions, followed by flow cytometry to detect microdroplets containing microcolonies. This allows previously uncultured organisms to be grown in pure culture to allow studies of microbial physiology and metabolism [54]. The advantages of culture and identification of these new strains is two fold. Firstly they will allow a greater understanding of how these organisms function within their ecosystem and therefore the processes they may drive on this ecosystem scale. Secondly they provide a new reservoir of organisms that can be screened for new traits, secondary metabolites and other metabolic products for potential applications.

5.2 Molecular based

Molecular-based approaches are now being applied to link uncultured microbes with their specific function and ecological properties, building on from the novel culture-based studies mentioned above. Environmental genomics provides one such approach, and involves sequencing large chunks of DNA recovered from a specific habitat to provide information on species richness, physiological activities, metabolic pathways and ecological strategies in such a habitat. This approach is aided by the observations that bacterial genomes are organized in clusters of linked genes (operons, pathogenicity islands or fitness islands) that provide adaptive properties to the cell. Béjà *et al.* [6], used the genomic approach to identify an abundant, but as yet uncultured marine *Proteobacterium* in the sea that contained the gene encoding for proteorhodopsin. Previously the gene was only known in extreme halophilic *Archaea*, and this purely genomic approach discovered a new and significant type of phototrophy in the oceans driven by bacteria. These methods provide a genomics perspective to microbial population biology, ecology and genome evolution. Recent large environmental metagenomic sequencing projects such as the sequencing of the Sargasso Sea, are providing an enormous database for discovering the novel prokaryotic diversity and their associated metabolic capabilities within this habitat [50]. The challenge however is integrating all this information to elucidate the role of the microorganisms in the total ecosystem [2].

The genomic approach to accessing microbial biodiversity can also allow the discovery of novel chemical structures, natural products, metabolic pathways, biochemistries and physiologies. Cloned DNA fragments isolated directly from environmental samples such as soil are easily exploitable in appropriate vectors and expression in surrogate systems for industrial scale production. Wang *et al.* [53] utilised this approach for discovery of new antimicrobial compounds, the terragines from soil-derived recombinant DNA libraries expressed in *Streptomyces lividans*. Similarly, Gillespie *et al.* [25] identified two broad-spectrum antibiotics, turbomycin A and B from a soil metagenomic library constructed and expressed in *Escherichia coli*. Biopanning of environmental libraries, in which gene probes are designed for a desired class of enzymes based on known sequences, and screened against environmental libraries, allows a further method to enrich for a particular gene class. For example, Courtois *et al.* [16] screened an environmental DNA library by PCR to identified new polyketide synthase genes which expressed several novel compounds, and demonstrated the potential of these methods as an effective drug discovery tool.

6. Policy for Australian Bio-discovery

The importance of biodiversity including its potential to spawn biodiscoveries and underpin new sustainable development, was recognised globally when the Convention on Biological Diversity was opened for signature at the Earth Summit in Rio de Janeiro in 1992. This legally binding international treaty now includes 187 countries as parties, including Australia. By ratifying the CBD, Australia committed itself to taking appropriate national and international measures to achieve the convention's three principle objectives: the conservation of biodiversity; the sustainable use of its components; and the equitable sharing of benefits arising out of the utilization of genetic resources [11].

However, until recently, Australia was not alone in its inability to deliver the CBD's objectives [9]. There was widespread acknowledgement that a lack of adequate legislative framework for procedural transparency and legal certainty for users of biodiversity, was hindering biodiscovery effort and facilitating biopiracy [9, 49]. The issues of access and benefit sharing continue to dominate the CBD agenda [11], which in 2002 produced 'The Bonn Guidelines' to specifically deal with access and benefit sharing [10].

At many levels, Australia has taken up this challenge and made important progress. There have been inquiries and reviews [51, 29], and some jurisdictions including Queensland, Western Australia and the Northern Territory have released policy papers for discussion and formal submission. To date, these processes have resulted in draft access and benefit sharing legislation for Commonwealth areas and Queensland (draft Environment Protection and Biodiversity Conservation Amendment Regulations 2004, Biodiscovery Bill Exposure Draft 2003), and new draft legislation is anticipated in other jurisdictions. The Commonwealth and the states and territories have also entered an Intergovernmental Agreement to endorse a nationally consistent approach to implement the 'Bonn Guidelines' and facilitate biodiscovery through application of existing laws and development of new ones [19].

Consistent with the CBD, this emerging Australian access and benefit sharing framework includes within its scope Australia's microbial biodiversity and its associated immense biodiscovery opportunity. However, this framework does not protect these culturable and unculturable microbe resources in international collaborations that require movement of material offshore. This is because export regulations under the Environment Protection and Biodiversity Conservation (EPBC) Act currently exempt all prokaryotes and viruses from the application of the regulations. This loophole means that even if a conditional access permit is obtained for collection of source material, microbial material, including in viable state, can be exported offshore without restriction. Where collections and research are funded internationally, this often results in significant loss of intellectual property (IP) associated with the collected material and data pertaining to it. Even if funds are available for basic microbiological research, the lack of a consistent policy on access and benefit sharing inhibits bona fide researchers from progressing discovery and development of biotechnological leads within Australia.

This lack of control over non-eukaryotic microorganisms lies in stark contrast to the tight export regulations for macroorganisms, and there is great irony in this dichotomy. The biodiversity of microorganisms, and hence biodiscovery potential, is believed to far outstrip that of macroorganisms. Further, compared to the finite amounts of un-reproducible raw sample or extract of macroorganisms required for development of biodiscovery leads, transfer of viable strains of microorganisms without regulation, often selectively isolated after considerable Australian R&D effort, also gives away the long-term large-scale supply method.

The issue of inadequate regulation of microbial exports is recognised [51, 29] and there are opportunities to tighten this loophole through appropriate lobbying and support. There is now an urgent need for policy and legislative development to occur in parallel with microbial research in Australia, so that research including biodiscovery can be facilitated while maximising benefits to Australia and ensuring environmental sustainability. The policy and legislative setting is still in a formative and consultative phase in most Australian jurisdictions, so the opportunity exists for researchers and scientists to interact with policy makers and legislators to achieve this outcome.

7. Conclusions

We live at a time of unprecedented access to new tools allowing examination of microorganisms, even previously unculturables. The challenge ahead is to facilitate and embrace this revolution and maximise and distribute the benefits it offers. Access to this technology should not come at a price of trade off of IP, increase the value of the IP generated and loss of cultures, but rather must enhance ownership of IP and better utilise data for both biodiscovery and biodiversity/conservation outcomes.

References

1. Ashton, R., Burnell, J., Furbank, R., Jenkins, C. and Hatch, M. (1990). Enzymes of C4 photosynthesis. *Methods in Plant Biochemistry,* **3**:39-71.
2. Azam, F. and Worden, A. Z. (2004). Microbes, molecules and marine ecosystems. *Science,* **303**: 1622-1624.
3. Barnes, R.D. (1987). *Invertebrate Zoology*; 5th Edition. Fort Worth, TX: Harcourt Brace Jovanovich College Publishers. pp.92-96, 127-134, 149-162.
4. Barnes, R. S. K. and Hughes, R. N. (Eds.) (1999). *An Introduction to Marine Ecology.* 3rd edition. Oxford, UK: Blackwell Science Ltd. pp.117-141.
5. Béjà, O., Suzuki, M.T., Heidelberg, J.F., Nelson, W.C., Preston, C.M., Hamada, T., Eisen, J.A., Fraser, C.M. and DeLong, E.F. (2002). Unsuspected diversity among marine aerobic anoxygenic phototrophs. *Nature,* **415**: 630-633.

6. Béjà, O., Aravind, L., Koonin, E.V., Susuki, M.T., Hadd, A., Nguyen, L.P., Jovanovich, S.B., Gates, C.M., Feldman, R.A., Spudich, J.L., Spudich, E.N. and DeLong, E.F. (2000). Bacterial rhodopsin: evidence for a new type of phototrophy in the sea. *Science,* **289**: 1902-1906.
7. Binder S., Levitt A. M., Sacks J. J. and Hughes J. M. (1999) Emerging infectious diseases: public health issues for the 21st century. *Science*, **284**: 1311-1313.
8. Button, D. K., Schut, F., Quang, P., Martin, R. and Robertson, B. R. (1993). Viability and isolation of marine bacteria by dilution culture: theory, procedures and initial results. *Applied and Environmental Microbiology*, **59**: 881-891.
9. CBD (Convention on Biological Diversity), (2000). Report of the Panel of Experts on Access and Benefit Sharing presented to the 5th Conference of the Parties.
10. CBD (Convention on Biological Diversity) Secretariat, (2002). Bonn Guidelines on Access to Genetic Resources and Fair and Equitable Sharing of the Benefits Arising out of their Utilization. UNEP, Canada. ISBN: 92-807-2255-7.
11. CBD (Convention on Biological Diversity) Secretariat, (2003). Handbook of the Convention on Biological Diversity, 2nd Edition, UNEP, Canada. ISBN: 92-807-2280-8.
12. Cho, J.-C. and Giovannoni, S. J. (2004). Cultivation and growth characteristics of a diverse group of oligotrophic marine gammaproteobacteria. *Applied and Environmental Microbiology*, **70**: 432-440.
13. Cohen M. L. (2000). Changing patterns of infectious disease. *Nature*, **406**: 762-767.
14. Connon, S. A. and Giovannoni, S. J. (2002). High-throughput methods for culturing microorganisms in very-low-nutrient media yield diverse new marine isolates. *Applied and Environmental Microbiology*, **68**: 3878-3885.
15. Cook, T. (2002). Microbial biodiversity: saving bacteria to save ourselves. *Harvard Science Review*, Winter 2002: 26-28.
16. Courtois, S., Cappellano, C. M., Ball, M., Francou, F.-X., Normand, P., Helynck, G., Martinez, A., Kolvek, S. J., Hopke, J., Osburne, M., August, P. R., Nalin, R., Guérineau, M., Jeannin, P., Simonet, P. and Pernodet, J. L. (2003). Recombinant environmental libraries provide access to microbial diversity for drug discovery from natural products. *Applied and Environmental Microbiology*, **69**: 49-55.
17. Curtis, T. P., Sloan, W. T. and Scannell, J. W. (2002). Estimating prokaryotic diversity and its limits. *Proceedings of the National Academy of Science*, **99**: 10494-10499.
18. Davidson, S.K., Allen, S. W., Lim., G. E., Anderson, C. M. and Haygood, M. G. (2001). Evidence for the biosynthesis of bryostatins by the bacterial symbiont "*Candidatus Endobugula sertula*" of the bryozoan *Bugula neritina*. *Applied and Environmental Microbiology*, **67**: 4531-4537.
19. DEH (Department of Environment and Heritage), (2002). *Nationally consistent Approach for Access to and the Utilisation of Australia's Native Genetic and Biochemical Resources.* Commonwealth of Australia, Canberra.
20. Eilers, H., Pernthaler, J., Glöckner, F. O. and Amann, R. (2000). Culturability and in situ abundance of pelagic bacteria from the North Sea. *Applied and Environmental Microbiology*, **66**: 3044-3051.

21. Errampalli, D., Leung, K., Cassidy, M.B., Kostrzynska, M., Blears, M., Lee and H. and Trevors, J.T. (1999). Applications of the green fluorescent protein as a molecular marker in environmental microorganisms. *Journal of Microbiological Methods,* **35**: 187-199.

22. Fenical, W. (1997). New Pharmaceuticals from marine organisms. *Trends in Biotechnology,* **15**: 339-341.

23. Fuhrman, J. A. (1999). Marine viruses and their biogeochemical and ecological effects. *Nature,* **399**: 541-548.

24. Fuhrman, J. A. and Davis, A. A. (1997). A widespread archaea and novel bacteria from deep sea as shown by 16S rRNA gene sequences. *Marine Ecology Program Series,* **150**: 275-285.

25. Gillespie, D. E., Brady, S. F., Bettermann, A. D., Cianciotto, N. P., Liles, M. R., Rondon, M. R., Clardy, J., Goodman, R. M. and Handelsman, J. (2002). Isolation of antibiotics turbomycin A and B from a metagenomic library of soil microbial DNA. *Applied and Environmental Microbiology,* **68**: 4301-4306.

26. Haefner, B. (2003). Drugs from the deep: Marine natural products as drug candidates. *Drug Discovery Today,* **8**: 536-544.

27. Holm, L. G., Plucknett, D. L., Pancho, J. V. and Herberger, J. P. (1997). *The world's worst weeds, distribution and biology*. University Press, Hawaii.

28. Horn, C., Schmid, B.G., Pogoda, F.S. and Wimmer, E.A. (2002). Fluorescent transformation markers for insect transgenesis. *Insect Biochemistry and Molecular Biology,* **32**: 1221-1235.

29. HOR SCPIRS (House of Representatives Standing Committee on Primary Industries and Regional Services), (2001). *Bioprospecting: Discoveries Changing the Future*. The Commonwealth of Australia, Canberra. ISBN: 0642 46359 X.

30. Kaeberlein, T., Lewis, K., and Epstein, S. S. (2002). Isolating "Uncultivable" microorganisms in pure culture is a simulated natural environment. *Science,* **296**: 1227-1129.

31. Karner, M. B., DeLong, E. F. and Karl, D. M. (2001). Archaeal dominance in the mesopelagic zone of the pacific ocean. *Nature,* **409**: 507-510.

32. Keller, M. and Zengler, K. (2004). Tapping into microbial diversity. *Nature Reviews, Microbiology,* **2**:141-150.

33. Kim, J. and Park, E. J. (2002). Cytotoxic anticancer candidates from natural resources. *Current Medicine and Chemistry Anti-Cancer Agents,* **2**: 485-537.

34. Lalli, C.M. and Parsons, T. R. (1995). Biological Oceanography: An Introduction. Butterworth-Heinemann Ltd. Oxford, UK. pp.220-233.

35. Levinton, J.S. (1995). *Marine Biology: Function, Biodiversity, Ecology*. Oxford University Press, Inc. New York. pp.306-319.

36. Llewellyn, L. E. and Burnell, J. N. (2000). Marine organisms as sources of C_4-weed-specific herbicides. *Pesticide Outlook,* **11(2)**: 64-67.

37. Massana, R., Delong, E. F. and Pedros-Alio, C. (2000). A few cosmopolitan phylotypes dominate planktonic archeal assemblages in widely different oceanic provinces. *Applied and Environmental Microbiology,* **66**: 1777-1787.

38. Neu, H. C. (1992). The crisis in antibiotic resistance. *Science,* 257: 1064-1073.

39. Newman, D. J., Cragg, G.M. and Snader, K.M. (2003). Natural products as sources of new drugs over the period 1981-2002. *Journal of Natural Products*, **66**: 1022-1037.

40. Newman, D. J., Jensen, P. R., Clementi, J. J. and Acebal, C. (1989). Novel activities from marine-derived microorganisms. In *Novel microbial products for medicine and agriculture,* (Demain, L.A., G. A. Somkuti, and J. C. Hunter-Cevera, Eds.). Society for Industrial Microbiology, Amsterdam, The Netherlands. pp.239-251.

41. Palmieri, S., Sebastio, L., Mele, G., Annunziata, M., Annunziata, S., Copia, C., Viola, A., De Simone, M., Pocali, B., Schiavone, E.M. and Ferrara, F. (2002). High-dose cytarabine as consolidation treatment for patients with acute myeloid leukemia with t(8;21). *Leukemia Research,* **26**: 539-543.

42. Rappé, M.S. and Giovannoni, S. J. (2003) The uncultured microbial majority. *Annual Review of Microbiology,* **57**: 369-394.

43. Rohwer, F., Seguritan, V., Azam, F. and Knowlton, N. (2002). Diversity and distribution of coral-associated bacteria. *Marine Ecology Progress Series,* **243**:1-10.

44. Schwartsmann, G., Brondani da Rocha, A., Berlinck, R. G., Jimeno, J. (2001). Marine organisms as a source of new anticancer agents. *The Lancet Oncology,* **2**: 221-225.

45. Sherr, E. and Sherr, B. (2000). Marine Microbes: An overview. In *Microbial Ecology of the Oceans* (Kirchman, D. L., Ed.). Wiley-Liss, Inc. New York, US. pp.13-46.

46. Shiba, T., Simidu, U. and Taga, N. (1979). Distribution of Aerobic bacteria which contain Bacteriochlorophyll *a. Applied and Environmenatal Microbiology,* **38**: 43-45.

47. Stewart Jr, C. (2001). The utility of groon fluorescent protein in transgenic plants. *Plant Cell Reproduction,* **20**: 376-382.

48. Sumich, J.L. (1996). An Introduction to the Biology of Marine Life, 6th edition. Wm. C. Brown. Publishers, Dubuque, IA. pp.255-269.

49. Ten Kate, K. and Laird, S. A. (1999). *The Commercial Use of Biodiversity – Access to Genetic Resources and Benefit Sharing.* Earthscan Publications, London. ISBN: 1 85383 334 7.

50. Venter, J. C., Remington, K., Heidelberg, J. F., Halpern, A. L., Rusch, D., Eisen, J. A., Wu, D., Paulsen, I., Nelson, K. E., Nelson, W., Fouts, D. E., Levy, S., Knap, A. H., Lomas, M. W., Nealson, K., White, O., Peterson, J., Hoffman, J., Parsons, R., Baden-Tillson, H., Pfannkoch, C., Rogers, Y-H and Smith, H. O. (2004). Environmental genome shotgun sequencing of the Sargasso Sea. *Science,* Published online March 4 2004; 10.1126/ science. 1093857 (Science Express Research Articles)

51. Voumard, J. (2000). *Access to Biological Resources in Commonwealth Areas.* Commonwealth of Australia, Canberra. ISBN: 0642547157.

52. Walsh, C. (2000). Molecular mechanisms that confer antibacterial drug resistance. *Nature,* **406**: 775-781.

53. Wang, G.-Y.-S., Graziani, E., Waters, B., Pan, W., Li, X., McDermott, J., Meurer, G., Saxena, G., Andersen, R. J. and Davies, J. (2000). Novel natural products from soil DNA libraries in a Streptomycete host. *Organic Letters,* **2**: 2401-2404.

54. Zengler, K., Toledo, G., Rappé, M., Elkins, J. Mathur, E. J., Short and J. M., Keller, M. (2002). Cultivating the uncultured. *Proceedings of the National Academy of Science,* **99**: 15681-15686.

Fifty years of selection on *Streptogramin* producing strains

F. Gosselé and M. Piecq

Phibro Animal Health, Rue de l'Institut 87A, B-1330 Rixensart, Belgium

1. Introduction

Numerous antibiotic producing microorganisms have been isolated and described since the discovery of penicillin by Fleming. Pharmaceutical companies started important strain improvement programs on these organisms once the clinical and commercial interest of the new molecules was confirmed. Strain improvement traditionally focuses on quantitative aspects of the fermentation with production titer increase as the number one criteria. However, certain approaches in strain improvement may lead to the isolation of mutants that accumulate precursors or shunt products that can be ultimately used for the development of new semi-synthetic derivatives. The objective of this overview is to illustrate these quantitative and qualitative developments using two streptogramin producing actinomycetes, *Streptomyces virginiae* and *Streptomyces pristinaespiralis.*

2. *Streptomyces virginiae 899* and *virginiamycin*

In 1954 De Somer and Van Dijck [11] isolated from soil a *Streptomyces* strain producing a new antibiotic. Both the strain and its bioactive compound were labeled 899. Physicochemical and biological investigations on the compound revealed strong similarity to the newly discovered streptogramin [7]. The isolate was identified as *Streptomyces virginiae*, described in *Bergey's Manual of Determinative Bacteriology* (1948), based on its morphological characteristics on different media as well as its biochemical properties. The strain was deposited to ATCC under the access number ATCC 13161. Two different names were later given to antibiotic 899: staphylomycin and, finally, virginiamycin.

Excellent mouse protection by virginiamycin was recorded and experimental infections with *Streptococcus pyogenes* were successfully treated with partially purified virginiamycin [11]. Later, Eyssen *et al*. [12] reported the growth promoting action of virginiamycin in chickens, explained by its effect on the crop microflora.

Virginiamycin was then further developed as a growth promoter for chicken, swine and cattle under the commercial name Stafac®. This interest led rapidly to the development of small scale experiments to obtain larger quantities of the compound. Van Dijck *et al*. [22] reported the first development work on production media for shake flask testing. Virginiamycin was assayed by plate diffusion with *Staphylococcus aureus* ATCC 6538 as indicator organism. The authors also described optimal shaking conditions of the Erlenmeyer flasks and mutagenesis protocols using U.V. or mustard gas were given. Natural selection as well as mutagenesis resulted in the isolation of colonies with variable diameter that could be classified in different groups: deep red, white, bald and slightly yellow. A general correlation between morphological traits and virginiamycin yield could not be found.

This was the beginning of an important selection program on *Streptomyces virginiae* 899 yielding several hundreds of thousands of strains till now. The objectives of such type of strain development program are quantitative (i.e. directly economic) as well as qualitative. In the former category of objectives we find titer increase, maintained improved productivity during extended fermentation cycles (resulting in smaller effluent volumes to be evacuated), decreased consumption of expensive raw materials (e.g. glucose), decreased air consumption, lowered viscosity (resulting in lower energy costs or impacting downstream processing performance) and possibility to produce at higher temperature (lowering industrial fermentation cooling costs). Qualitative objectives are the isolation of strains with a modified impurity profile corresponding better to end product specifications (e.g. elimination of fermentation related impurities) and the isolation and characterization of blocked mutants. Both quantitative and qualitative aspects will be discussed further.

The first focus of strain development programs is on the isolation of high yielding mutants, a fundamental objective for the initial development of the *Streptomyces virginiae* industrial fermentation. The starting point in virginiamycin culture development was strain 899, with low yields, sensitive to its own antibiotic and, as it was discovered later, sensitive to bacteriophage [4]. Hence, initial selection work targeted these three weaknesses. Mutagenesis techniques applied were mainly U.V. irradiation, nitrogen mustard [4] and, later, γ-radiation and N-methyl-N'-nitro-N-nitroso-guanidine (MNNG).

Both numerical (i.e. random) and rational approaches were followed. Traditional numerical screening in the 50-60ies consisted of testing a large number of strains in Erlenmeyer flasks on shaking tables with their antibiotic concentration determined by a microbiological assay. The flow diagram with the different manual steps was the following:

Mutagenic treatment

↓

Surviving mutants on media in Petri dishes

↓

Transfer to agar slants

↓

Sporulation on slants

↓

Seed development in Erlenmeyer flask

↓

Production in Erlenmeyer flask

↓

Microbiological assay

Virginiamycin is extracted from fermentation broth with a solvent (methanol, acetone, isopropanol), diluted in buffer and tested in a bioassay with a sensitive organism, for example *Staphylococcus aureus, Micrococcus luteus* or *Corynebacterium xerosis* [6]. This assay requires a preliminary estimation of the expected production titer in order to prepare the appropriate dilution of the extract, ready for microbiological testing.

A “unit” research group dealing with this type of selection consists of three persons involved in medium preparation, mutagenesis, spore and seed transfers, antibiotic determination and analysis of the results. Such a unit system represents a testing capacity of 200 to 250 cultures per week.

For rational screening, a basic knowledge of antibiotic biosynthesis and microbiology or genetic background of the strain is required. Virginiamycin is composed of two factor groups, M and S, each composed of major (M_1, S_1) and minor components (M_2, M_3, S_4). Their structures are given in Fig. 1. Component M_1 is derived from

Figure 1. Structures of the virginiamycin M and S components

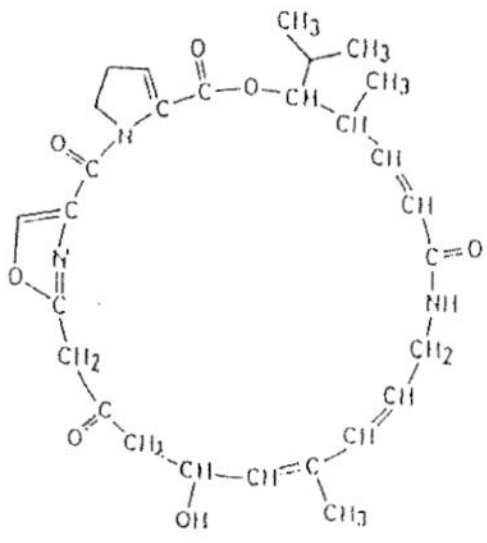

Factor M 1

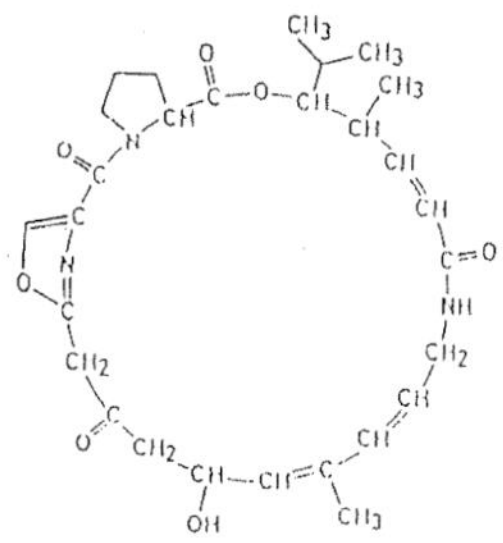

Factor M 2

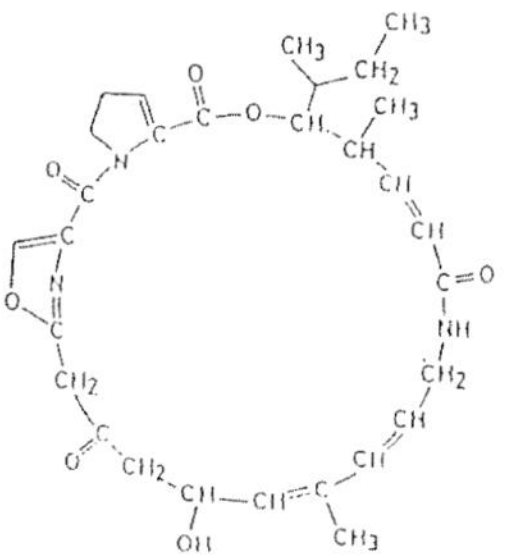

Factor M 3

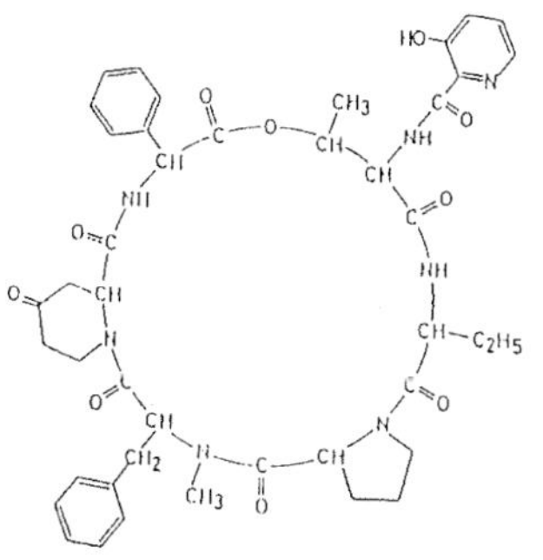

Factor S 1

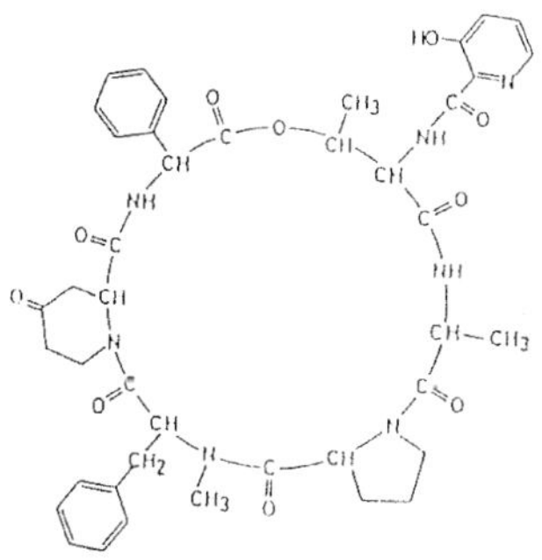

Factor S 4

valine, acetate, glycine, serine, methionine and proline [14]. It was found by double labeling experiments, using ^{14}C-proline and ^{3}H-acetate, that the less active component M_2 is an immediate precursor of M_1 [21]. The structure of the shunt product M_3 was elucidated later (Kolpak and Gosselé, unpublished). Components S_1 and S_4 are non ribosomal peptides and have in common the amino acids threonine, proline and phenylalanine. The 4-oxo-pipecolic acid and 3–hydroxypicolinic acid moieties of both components are derived from lysine [17]. S_4 and S_1 differ in one amino acid: 2-aminobutyric acid is incorporated in S_1; S_4 contains alanine. The nomenclature of virginiamycin factors and those of other streptogramins was reviewed by Crooy and De Neys [10].

A second characteristic, typical for the members of the streptogramin family of antibiotics and important for assay development is the synergy effect. Factors M and S do not only have an additive effect: their antibacterial effect is multiplied by a factor of synergy, depending on the ratio of M and S concentrations. The molecular mechanism governing this synergy was reviewed by Cocito [9]. The maximum synergy factor of 2.7 is obtained at an M/S ratio of 65/35 [6]. The original *Streptomyces virginiae* strain 899 produces virginiamycin with an M/S ratio below optimum, i.e. 80/20.

The numerical approach, combined with rational screening, allowed the isolation of the first interesting improved mutants. Highlights in this initial stage in the strain development on *Streptomyces virginiae* 899 were: strains with increased resistance to virginiamycin, strains cured from their bacteriophages, resistants to amino acid analogs, and revertants of auxotrophs. This resulted in a 1000-fold increase of the production titer [4]. It is remarkable that all these mutants, in spite of their obvious improvement, conserved the suboptimal M/S ratio of 80/20.

The possibilities to increase the capacity of such numerical screens with microbiological assays becomes limited once the production titers reach industrially interesting levels and when efficient mutagenesis techniques yield very different strains. As mentioned above, some estimation of the expected titer is necessary to adapt the extract dilution to fit with the sensitivity of the target organism (in the case of *Micrococcus luteus,* this target concentration is 1 μg / ml). With high and variable titers, this estimation becomes difficult. New analytical techniques were needed. A reversed phase HPLC assay with U.V. detection at 214 nm was developed based on the method of Gosselé *et al.* [13]. Virginiamcyin is extracted from fermentation broth in aqueous isopropanol, diluted in mobile phase and injected. The different M and S components are quantified and the resulting synergistic activity is calculated based on the ratio of M_1/S_1. Only these two components are taken into account since they are quantitatively and qualitatively the most important ones. Partial automation of the HPLC system is possible and selection capacity could be increased up to approximately 1000 strains per week with a "unit" research group.

The HPLC assay also allowed to easily detect special mutants: M only producers (M+S-), S only producers (M-S+), non producers (M-S-) and strains blocked in their M_2 to M_1 transformation capacity. S components and some of their precursors are fluorescent: this is due to the 3-hydroxypicolinic acid moiety [8]. By using a double detection method coupled to the HPLC, i.e. U.V.- and fluorescence, specific mutants blocked in their S biosynthesis and thus accumulating fluorescent intermediates could be isolated. These blocked mutants are interesting for several reasons: their revertants are often improved producers, they are useful strains to study the biosynthesis and, last but not least, blocked mutants sometimes accumulate precursors that can be used for the preparation of semi-synthetic molecules. Fermentation media inoculated with these mutants were extracted with methyl-isobutylketone (MIBK); these extracts were concentrated by distillation and the fluorescent compounds were precipitated with n-hexane. Further purification was by adsorption column chromatography, re-extraction with MIBK and re-precipitation with n-hexane. Products thus obtained were identified by NMR. Fluorescent "S_1 and S_4-like" compounds lacking the 4-oxopipecolic acid and the N-methyl phenylalanine moiety either in closed or in open structure were thus isolated and identified [Gosselé, Marneffe and Kolpak, unpublished results]. It was later found that many strains derived from *Streptomyces virginiae* 899 produce minute amounts of these compounds, but only the blocked mutants accumulated them in reasonable quantities.

With the microbiological approach these blocked mutants would be found to be poor or non producers. Only an "experienced eye" would in some cases be capable to pick up special strains from the microbiological test, based on the morphology of aberrant inhibition zones. The chance to pick them up was considerably increased by the use of HPLC methods.

Space limitation can at a certain moment become a bottleneck to further capacity increase. Miniaturization, a third stage in the history of *Streptomyces virginiae* strain development, was the answer to this. The Erlenmeyer flasks were replaced by microtiter plates with 24 wells [19]. Theoretically, this seems straightforward but designing such miniaturized and automated systems requires a stepwise development. The parameters to be determined are medium volume (for seeds as well as production), inoculum volume, agitation conditions (rpm, stroke), humidity and stacking of the microtiter plates (since aeration is influenced by the way the plates are stacked in the shaking boxes). In miniaturized systems, the duration of the fermentation is limited due to the evaporation. Researchers are tempted to use the best fermentation conditions to obtain high titers in the miniaturized system but the question raises whether, for strain selection purposes, this is the right approach. Indeed, screening systems should have in the first place a high discriminatory power to identify rare improved mutants. Increasing the average titers of the whole population is not an objective on its own.

Omitting time consuming steps from the previous strain selection scheme was the next critical step. The sporulation stage on slants was eliminated and replaced by direct inoculation of the mutants in the seed medium in wells. The scheme then becomes:

Mutagenic treatment

↓

Mutant colonies on Petri dishes

↓

Seed development in microtiter plate

↓

Production in microtiter plate

↓

HPLC assay

Medium distribution, seed transfer and broth extraction, followed by HPLC analysis can be automated. Although appropriate robotics with integrated image analysis software exist, the inoculation of colonies in seed medium is still done manually since we estimate that human judgment on the type of colonies to be picked up may either avoid the useless testing of recognizable non-producers or favor the testing of rare mutants. After the inoculation of the production media, the seeds are conserved at 4 °C. Once the HPLC results are available, the seeds of the interesting mutants are streaked on sporulation slants. These sporulated cultures are used for the preparation of working stocks to be tested later at pilot scale.

In the late nineties we introduced a fourth stage in the strain development strategy: the use of Near Infra Red (NIR) analysis to estimate virginiamycin production titers in broth [19]. NIR analysis does not quantify virginiamycin but correlates variation within the spectra (Fig. 2) with variation of the virginiamycin concentrations according to a pre-established model.

This model is developed by correlation between a “calibration” set of spectra and their corresponding production titers determined by HPLC (Fig. 3).

Routine NIR analysis does not require sample preparation. Fermentation broths are presented as such, without any treatment, automatically to the NIR analyzer. This

Figure 2: NIR spectra of a set of virginiamycin containing fermentation samples

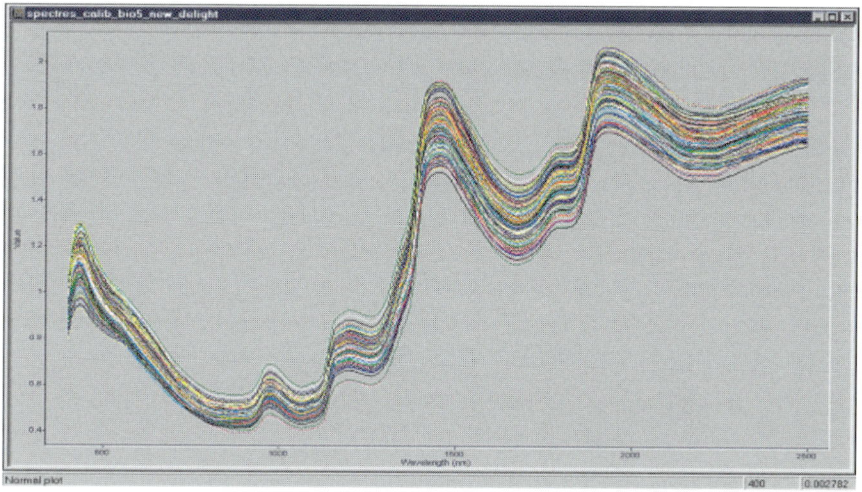

Figure 3: Correlation between virginiamycin titers determined by HPLC (X – axis) and predicted by NIR analysis (Y – axis)

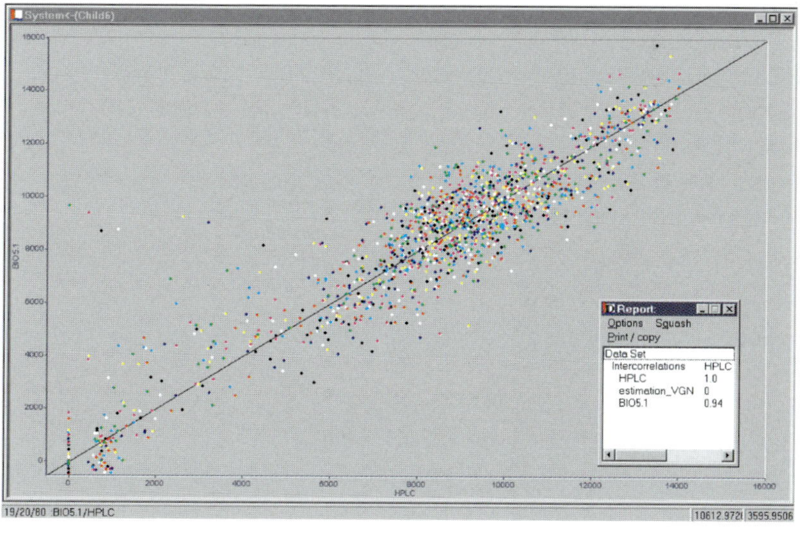

allows a significant strain selection capacity increase: up to 2000 mutants per week can be investigated by a "unit" group. Another advantage of NIR analysis is that a second or a third parameter (e.g. glucose consumption, biomass) can be followed simultaneously in the same analysis as long as the appropriate calibration models are available.

One has to live, however, with the fact that each medium change introduced with the purpose to work as closely as possible to the industrial environment, may require a new NIR calibration. NIR analysis is also less precise than the HPLC assay. As such this is not an issue since the strain selection programs expect production titer jumps greater than 10 % and since a second trial on the best mutants is anyway performed by HPLC.

Furthermore, the combination of NIR and HPLC analysis allows the isolation of new kinds of mutants. For instance, the supplementary use of a semi quantitative model (Principal Component Analysis) during the first trial focuses on aberrant NIR spectra. The second confirmatory experiment with the interesting isolates combines NIR and HPLC analysis and allows the identification of blocked mutants, the elimination of contaminated broths and the confirmation of special features of some mutants such as abnormal growth or modifications of their metabolic abilities.

NIR analysis of broths in these miniaturized systems allowed the isolation of a series of mutants with significant production titer increases. From morphological point of view, we observe that some of these interesting strains have a poor sporulation capacity and show a more fragmented mycelium. Some mutants had a modified carbohydrate utilization profile. It is remarkable that the M biosynthesis was oriented to a better transformation of M_2 into M_1. Strains also have a more favorable glucose conversion ratio. Finally, synergy of the M and S factors tend to a more favorable ratio around 68/32. Appropriate process modifications (media, running conditions) were thus needed to express the full potential at industrial scale.

The application of these different strain selection approaches in terms of mutagenesis as well as screening technology and analytics during almost fifty years resulted in the isolation of several hundreds of thousands of strains of which a small percentage is interesting for commercial (i.e the overproducers or strains with other characteristics important for quality) or scientific reasons (e.g. the blocked mutants). Several phenotypically different lineages are thus available. They differ in their productivity, pigmentation, sporulation capacity and utilization of carbon- and nitrogen sources.

Lanoot *et al.* [15] examined the phylogeny of a representative set of 20 phenotypically different strains from the *Streptomyces virginiae* 899 genealogy using current molecular biology tools, i.e. rep-PCR fingerprinting, DNA:DNA hybridizations and AFLP. The *Streptomyces virginiae* type strain (LMG 20534^T), the original 899 strain conserved at the ATCC collection (ATCC 13161) and the MAFF 116014 strain used to study the biosynthesis and effect of γ-butyrolactone autoregulator [23] were also included in this study.

DNA of strains 899 and MAFF 116014 have a low homology of 47 % and 54 % respectively with DNA of the *Streptomyces virginiae* type strain [1, 15], indicating that these strains do not belong to *Streptomyces virginiae*. The type strain and 899 had also clearly distinguishable rep-PCR patterns. On the other hand, ATCC 13161, MAFF 116014, strain 899 and all its mutants (overproducers as well as blocked mutants) showed identical rep-PCR patterns to each other. Therefore, rep-PCR proved to be an efficient technique to delineate genotypically the mutants derived from 899 and to demonstrate the similarity between 899 and MAFF 116014. The use of a finer technique like AFLP was necessary to distinguish between 899 and some of its non producing mutants, known to be deleted from large 20 kb DNA regions (Dehottay and Piecq, unpublished).

3. *Streptomyces pristinaespiralis*, *pristinamycin* and its semi-synthetic derivatives

The isolation of blocked mutants during the strain selection program is interesting for different purposes. As mentioned previously, generating revertants can result in improved yields. Secondly, blocked mutants are interesting study objects to investigate the antibiotic biosynthesis. Finally, they are possible starting points for new antimicrobials. This will be illustrated with the example of pristinamycin (originally called 7293 RP), another member of the streptogramin family of antibiotics, produced by *Streptomyces pristinaespiralis* 5647 [20].

Pristinamycin is composed of components II_A (identical to virginiamycin M_1), II_B (identical to M_2), I_A and I_B [10]. The latter two belong to the same group of cyclic hexadepsipeptides as the virginiamycin S components. Comparison with the virginiamycin S structures in Fig. 4 shows that the difference is located at the phenylalanine moiety. I_A and I_B contain a 4-dimethylamino phenylalanine and 4-methylamino –phenylalanine residue respectively instead of the phenylalanine in virginiamycin S_1.

Figure 4: Structures of pristinamycin I_A and I_B. and virginiamycin S_1

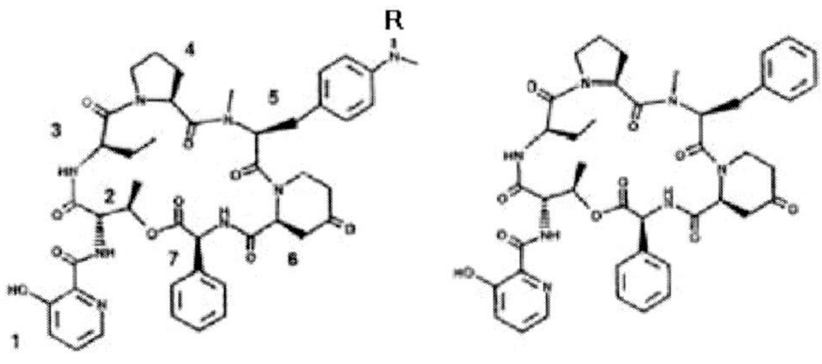

Pristinamycin I_A R= Me
Pristinamycin I_B R= H

Virginiamycin S_1

Pristinamycin is commercialized under the trade name Pyostacine. Initially it was used for treatment of staphylococcal infections associated with skin and bone infections [18]. It is active against the main bacteria causing respiratory diseases, i.e. *Staphylococcus aureus*, pneumococci and *Haemophilus influenzae* [16] and the agents of certain sexually transmissible diseases caused by *Neisseria, Mycoplasma* and *Chlamydia* [18]. However, the pristinamycin factors are poorly soluble in water and cannot be administered parentally. Hence, their interest to treat these clinically important infections is limited. Therefore, a series of water soluble derivatives of pristinamycin II and I were developed [3].

A first example is RP59500, later baptized Synercid and approved for the treatment of vancomycin resistant *Enterococcus faecium*. It is a mixture of dalfopristin (derived from the pristinamycin component II_A) and quinupristin (derived from pristinamycin I_A). The isolation of mutants of *Streptomyces pristinaespiralis*, producing only one of the pristinamycin factors, was disclosed by Barrere *et al*. [2]. Hence, fermentations with the individual blocked mutants could be optimized separately. This accelerated the development of the semisynthetic streptogramin by avoiding the highly sophisticated steps needed to separate and purify the components of the pristinamycin complex.

A second example is RP 106972, a new oral streptogramin, developed for the treatment of community-acquired respiratory tract, skin and genital infections [3]. It consists of two minor components of the pristinamycin complex, i.e. 45 % I_B (Fig. 4) and 55 % II_B (= virginiamycin factor M_2, see Fig. 1). The pristinamycin produced by the natural *Streptomyces pristinaespiralis* only contains 3 – 10 % of component I_B. Chemical – and biotransformation of the more abundant pristinamycin factor I_A allowed the production of I_B at a reasonable yield. The rational development of a specific mutant allowed cost-effective production of I_B. This strain had a mutation in the biosynthetic pathway of the 4-dimethylamino phenylalanine moiety: *pap*A, one of the four genes involved, was disrupted and the mutant thus obtained can be fed with different synthetic phenylalanine derivatives, in this case 4-methylaminophenylalanine [3]. This approach is called mutasynthesis. The second component in the co-crystalline mixture RP 106972 is $II_{B.}$ Wild type *Streptomyces pristinaespiralis* produces low concentrations of II_B. Possibilities to accumulate this component through process adaptation are limited: poor aeration for example results in unefficient II_A to II_B transformation, but also decreases their global yield. On the other hand, purification of II_B requires special techniques to separate it from the more abundant II_A. Blanc *et al.* [5] constructed by disruption of the snaA gene in *Streptomyces pristinaespiralis* a strain accumulating II_B (M_2).

Synercid and RP 106972 are two examples of the successful use of efficient rational strain development allowing the isolation of interesting blocked mutants.

4. Conclusion

Numerical and rational strain development efforts on *Streptomyces virginiae* and *Streptomyces pristinaespiralis* using the different approaches described are at the basis of the implementation of industrially viable fermentation processes. Moreover, the isolation and characterization of blocked mutants not only allowed better insight in the biosynthesis of the antibiotics but constituted also the starting blocks for the creation of new semi-synthetic streptogramins, e.g. Synercid and RP106972.

References

1. Anzai, Y., Okuda, T., and Watanabe, J. (1994). Application of the random amplified polymorphic DNA using the polymerase chain reaction for efficient elimination of duplicate strains in microbial screening. *Journal of Antibiotics*, **47**: 183-193.

2. Barrere, G., Jumel, J., Lacroix, P., Lehmann, C. and Sabatier, A. (1993). *Micro-organisms, preparation method therefore and uses thereof.* Patent WO 93/20182.

3. Barrière, J.C., Berthaud, N., Beyer, D., Dutka-Malen, S., Paris, J.M. and Desnottes, J.F. (1998). Recent developments in streptogramin research. *Currents in Pharmaceutical Design,* **4**: 155-180.

4. Biot, A. (1984). Virginiamycin: properties, biosynthesis and fermentation. In *Biotechnology of industrial antibiotics,* (Vandamme, E.J., Ed.). Marcel Dekker Inc.

5. Blanc, V., Lagneaux, D., Didier, P., Gil, P., Lacroix, P., and Crouzet, J. (1995). Cloning and analysis of structural genes from *Streptomyces pristinaespiralis* encoding enzymes involved in the conversion of pristinamcyin II_B to pristinamcyin II_A (PII_A) : PII_A synthase and NADH:Riboflavin 5'-phosphate oxidoreductase. *Journal of Bacteriology,* **177**: 5206-5214.

6. Boon, B. and Dewart, R. (1974). Methods for identification and assay of virginiamycin in animal feeds. *Analyst,* **99**: 19-25.

7. Charney, J., Fisher, W.P., Curran, C.; Maclowitz, R.A. and Tytell, A.A. (1953). Streptogramin, a new antibiotic. *Antibiotics Annual,* 171-173.

8. Clays, K., Di Giambattista, M., Persoons, A. and Engelborghs, Y. (1991). A fluorescence lifetime study of virginiamycin S using multifrequency phase fluorometry. *Biochemistry,* **30**: 7271-7276.

9. Cocito, C. (1979). Antibiotics of the virginiamycin family, inhibitors which contain synergistic components. *Microbiological Reviews,* **43**: 145-198.

10. Crooy, P.and De Neys, R. (1972). Virginiamycin: nomenclature. *Journal of Antibiotics,* **25**: 371-372.

11. De Somer, P. and Van Dijck, P. (1955). A preliminary report on antibiotic number 899, a streptogramin-like substance. *Antibiotics and Chemotherapy,* **11**: 632-639.

12. Eyssen, H., De Prins, V. and De Somer, P. (1962). The growth-promoting action of virginiamycin and its influence on the crop flora in chickens. *Poultry Science,* **41**: 227-233.

13. Gosselé, F., Blain, Ph., Marneffe, Th. and Biot, A. (1991). High-performance liquid chromatographic determination of virginiamycin in Stafac premixes and animal feeds. *Analyst,* **16**: 1373-1380.

14. Kingston, D.G.I., Kolpak, M.X., LeFevre, J.W. and Borup-Grochtmann, I. (1983). Biosynthesis of antibiotics of the virginiamycin family. 3. Biosynthesis of Virginiamcyin M_1. *Journal of American Chemical Society,* **105**: 5106-5110.

15. Lanoot, B., Vancanneyt, M., Gosselé, F., Piecq, M., Cnockaert, M.C. and Swings, J. (2003). Intra-species genotypic relationships among bioactive *Streptomyces virginiae* strains and mutants of the virginiamycin producing strain 899. *Abstract book ISBA'13*, Melbourne, Australia p.81.

16. Leclercq, R. (1999). La pristinamycine et les infections respiratoires de l'adulte. Activité in vitro de la prisitnamycine sur les germes respiratoires. Symposium. *La Presse Medicale,* **28**: 6-9.

17. Molinero, A.A., Kingston, D.G.I. and Reed, J.W. (1989). Biosynthesis of antibiotics of the virginiamycin family. 6. Biosynthesis of virginiamycin S_1. *Journal of Natural Products,* **52**: 99-108.

18. Paris, J.M., Barrière, J.C., Smith, C. and Bost, P.E. (1990). The chemistry of pristinamycins. *Recent Progress in the Chemical Synthesis of Antibiotics*, 183-248.

19. Piecq, M., De Maesschalck, E. and Gosselé, F. (2001). Industrial strain selection: miniaturised automated system for *Streptomyces virginiae. Abstract Book ISBA'12*, Vancouver, Canada p.216.

20. Preud' homme, J., Belloc, A., Charpentié, Y., and Tarridec, P. (1965). Un antibiotique formé de deux groupes de composants à synergie d' action: la pristinamycine. C.R. Acad. Sc. Paris, **260**: 1309-1312.

21. Purvis, M.B., LeFevre, J.W., Jones, V.L., Kingston, D.G.I., Biot, A.M. and Gosselé, F. (1989). Biosynthesis of antibiotics of the virginiamycin family. 8. Formation of the dehydroproline residue. *Journal of American Chemical Society,* **111**: 5931-5935.

22. Van Dijck, P.; De Somer, P.; Vanderhaeghe, H. and Parmentier, G. (1956). Mycologische gegevens over een streptogramine-gelijkend bestanddeel. *Agricultura,* **4**: 195-220.

23. Yamada, Y., Nihira, T. and Sakuda, S. (1997). Butyrolactone autoregulators, inducers of virginiamycin in *Streptomyces virginiae:* their structures, biosynthesis, receptor proteins, and induction of virginiamycin biosynthesis. In *Biotechnology of Antibiotics* (Strohl, W.R., Ed.). Marcel Dekker.

In search of new probiotics: Diversity, selection and exploitation of health-promoting organisms

R. Temmerman, L. Masco, G. Huys and J. Swings

Laboratory of Microbiology, Ghent University, K.L. Ledeganckstraat 35, B-9000 Ghent, Belgium

1. Introduction

Triggered by emerging socio-economical phenomena such as the aging society and the steady growth of the world's population, care and attention for human physical condition and health have become more prominent than ever. One of the many ways to achieve and maintain a 'good health' is through a well-balanced and healthy nutrition. In this regard, probiotic products designed for health promotion may fulfill an important role as part of the daily diet. Although the potential health-promoting properties of Lactic Acid Bacteria (LAB) were already reported at the turn of the 19th century [10, 29], it was not before the mid-1960s that the term probiotic was used for the first time by Lilley and Stillwell [25] to describe substances excreted by one microorganism, which stimulate the growth of another microorganism. Over the years, numerous other definitions have been proposed for probiotics, some of which were excessively long, driven by the discussions on the need for viability of the organism, the sites and modes of action, etc. The Food and Agriculture Organization of the United Nations (FAO) has recently proposed the following definition for a probiotic: "a live microorganism which, when administered in adequate amounts, confers a health benefit to the host" [34]. This definition certainly does not exclude further modifications, but it clearly reflects the probiotic concept as a health-promoting microorganism. Despite the steadily increasing probiotic market [41] and number of studies dedicated to probiotic research, significant efforts are needed to standardize the meaning of the term probiotic and on the characterization of organisms with promising probiotic characteristics. Although this chapter will mainly deal with probiotics for human use, it should be noted that probiotic organisms are also increasingly used to promote the health of food animals such as pigs, poultry and aquaculture species [35, 47].

Probiotic organisms may originate from humans, animals or plants. Although the FAO and the World Health Organization (WHO) have stated that there is adequate scientific evidence indicating the potential health benefits of probiotic foods and suggesting that specific LAB strains are safe for human use [34], probiotics should not be referred to as biotherapeutic agents [28]. As a result, very few probiotic products are available as registered pharmaceutical products. Because the probiotic concept has originated from ancient fermented milk products, the dairy matrix has long been the most frequently used carrier for probiotic organisms. However, modern production technologies have resulted in the development of a wide array of product types including tablets, powders, capsules or even coatings and liquid ampoules.

2. Diversity of probiotic microorganisms

According to the definitions proposed for a probiotic, a large variety of microbial species and genera are considered to have probiotic potential (Table 1) most of which belong to the LAB. LAB constitute a heterogeneous group of Gram-positive, non-sporulating, non-respiring cocci or rods, exhibiting a strictly fermentative metabolism with lactic acid as the key metabolite. The description of this group of bacteria, originally being based on morphological and physiological features, has frequently been debated due to recent phylogenetic studies. The core of LAB is composed of the genera *Lactobacillus*, *Leuconostoc, Pediococcus* and *Streptococcus* with taxonomical revisions and description of new genera leading to the following list of LAB genera: *Aerococcus, Alloiococcus, Carnobacterium, Dolosigranulum, Enterococcus, Globicatella, Lactobacillus, Lactococcus, Lactosphaera, Leuconostoc, Oenococcus, Pediococcus, Streptococcus, Tetragenococcus, Vagococcus* and *Weisella* [5]. Although phylogenetically not belonging to the LAB, also the genera *Bifidobacterium, Propionibacterium* and *Brevibacterium* are used in food industries because some strains display LAB-like properties.

Probiotic effects are strain specific, meaning that multiple strains of a certain species may have probiotic potential [23]. Table 2 summarizes the most widely used probiotic strains and their suppliers [39]. Despite the considerable number of potentially probiotic species (Table 1) and the high variety of commercially available probiotic products, the diversity of marketed probiotic strains appears to be relatively low (Table 2) as demonstrated for *Bifidobacterium*-containing probiotic products [20].

Table 1: Microbial species used as probiotics (Modified from [22])

Lactobacillus	***Bifidobacterium***	**Other LAB**
L. acidophilus	*B. adolescentis*	*Enterococcus faecalis*
L. amylovorus	*B. breve*	*Enterococcus faecium*
L. brevis	*B. bifidum*	*Lactococcus lactis* subsp. *lactis*
L. casei	*B. longum* biotype *longum*	*Leuconostoc mesenteroides*
L. crispatus	*B. longum* biotype *infantis*	*Pediococcus acidilactici* subsp. *acidilactici*
L. delbrueckii	*B. animalis* subsp. *lactis*	*Streptococcus thermophilus*
subsp. bulgaricus	*B. animalis* subsp. *animalis*	
L. fermentum	*L. gallinarum*	Non-LAB
L. gasseri	*Bacillus cereus*	
L. johnsonii	*Bacillus clausii* (labeled as *L. sporogenes*)	
L. paracasei	*Bacillus subtilis*	
L. plantarum	*Escherichia coli*	
L. reuteri	*Propionibacterium freundenreichii*	
L. rhamnosus	*Saccharomyces cerevisiae*	
L. salivarius	*Saccharomyces boulardii*	

Given the large number of LAB species known so far, it can be assumed that many potential probiotic strains are still to be discovered. In the development of human probiotics, strains belonging to the genera *Lactobacillus*, *Bifidobacterium* and *Streptococcus* have been frequently used, primarily because of the perception that they are autochthonous members of the intestinal microflora [19]. In addition, these bacteria have traditionally been used in the manufacturing of fermented dairy products and have the GRAS status. However, not all of the new probiotic strains are necessarily of human origin or may not have a documented history of safe use. Consequently, before 'new' strains can be included into a probiotic product and

Table 2: Common probiotic strains and their supplier [33]

Strain*	Supplier
B. animalis subsp. *lactis* Bb-12	Chr. Hansen, Copenhagen, Denmark
B. longum BB536	Morinaga Milk Industry Co., Zama-City, Japan
B. longum SBT-2928	Snow Brand Milk Products Co., Tokyo, Japan
L. acidophilus DDS-1	Nebraska Cultures, Lincoln, USA
L. acidophilus NCFMä	Rhodia, Madison, WI, USA
L. acidophilus SBT-2062	Snow Brand Milk Products Co., Tokyo, Japan
L. casei DN114 001	Danone, Paris, France
L. casei Shirota	Yakult, Tokyo, Japan
L. fermentum RC-14	Urex Biotech, London, Canada
L. johnsonii LA1	Nestlé, Lausanne, Switserland
L. paracasei CRL 431	Chr. Hansen, Milwaukee, USA
L. plantarum 299V	Probi, Lund, Sweden
L. reuteri SD2112	McNeil Consumers, USA
L. rhamnosus DR20	New Zealand Dairy Board, Palmerston North, N.Z.
L. rhamnosus GG	Valio, Finland
L. rhamnosus GR-1	Urex Biotech, London, Canada
L. rhamnosus LB21	Essum AB, Umea, Sweden
L. salivarius UCC118	University College Cork, Ireland

* *L. Lactobacillus*; *B. Bifidobacterium*

become commercially available, extensive research has to be performed concerning the safety [26], functionality, health benefits and technological properties of the strain.

3. Selection of new probiotic strains

Whereas traditional starter cultures used in the dairy industry are selected for their ability to rapidly produce desirable organoleptic qualities of cultured dairy products, probiotic bacteria are selected for the potential to provide specific health or nutritional benefits following consumption. Such a selection involves many different criteria including safety, technological and functional aspects (Figure 1) [36].

Figure 1: Criteria used for screening and validation of new potential probiotic strains (modified from [34])

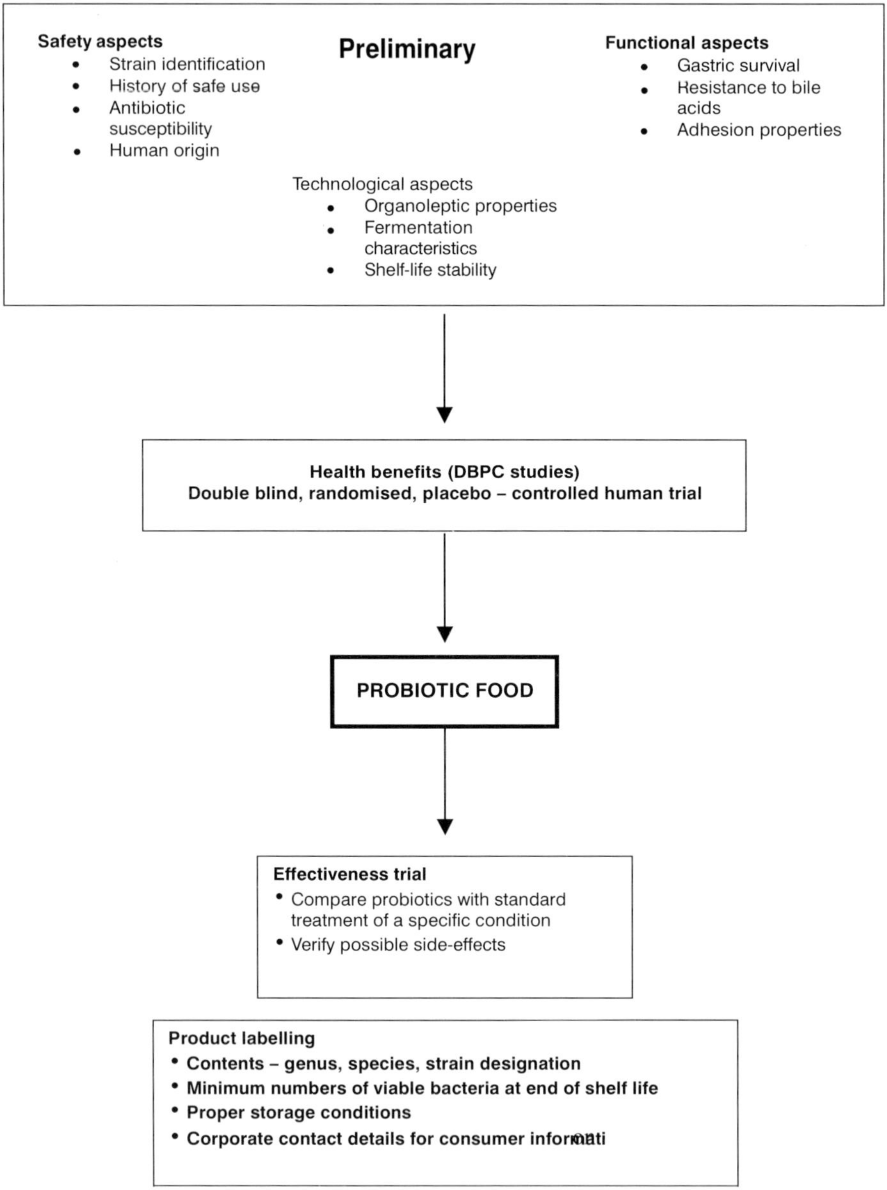

3.1. Origin of candidate probiotic strains

Although still under debate, it is suggested that a probiotic strain should preferably be a normal inhabitant of the intestinal tract, taking into account the possibility that many of these bacteria exhibit host specificity [36]. Therefore, if a strain is to be used as a probiotic for humans, it is highly desirable that the organism originates from a healthy human Gastro-Intestinal (GI-) tract. Alternatively, the search for new probiotic strains could be targeted at existing collections of micro-organisms such as international public culture collections or (private) research collections.

Before a strain can be considered for probiotic use, a large number of *in vitro* and *in vivo* characteristics have to be documented. First of all, the strain has to be thoroughly screened for its safety. Guidelines for the safety assessment have been addressed in several opinion articles [2, 11, 38].

3.2. Safety aspects of probiotic strains

The prerequisite of microbiological safety of a probiotic organism is the correct identification of the strain. Probiotic effects are strain specific, meaning that a correct characterization to the strain level is often necessary in order to link a strain to a specific health effect or application as well as to enable meaningful surveillance and epidemiological studies [13]. Species identification of probiotic organisms must be established using validated methodologies such as 16S rDNA sequencing, DNA fingerprinting or species-specific PCR tests [44].

Bacterial species that have a history of safe use for producing fermented dairy foods are generally considered as 'food grade' or GRAS organisms [22]. Historically, lactobacilli and bifidobacteria associated with food have been considered to be safe [1]. Their occurrence as normal commensals of the mammalian flora and their established safe use in a diversity of foods and supplement products worldwide support this conclusion. However, depending on their origin, probiotic strains may not always have this history of safe use and may theoretically be responsible for four types of side effects: systemic infections, deleterious metabolic activities, excessive immune stimulation in susceptible individuals, and (antibiotic resistance) gene transfer. In this regard, *in vitro* tests are crucial to verify several safety aspects of potential probiotic strains, for instance in order to determine the degree of enzymatic activity (bile salt deconjugation) or to test if a strain produces a toxic compound as indicated by the EU Scientific Committee on Animal Nutrition [40]. SCAN guidelines mainly apply to animal nutrition, although it has been suggested that for some aspects a rigorous extrapolation can be made to human nutrition.

Due to the indiscriminate use of antibiotics in human and veterinary medicine and in animal growth promotion, antibiotic resistance has become an increasingly common characteristic in (food-borne) microorganisms [45], causing serious problems in treatment of microbial infections. Antibiotic resistance in bacteria may be either intrinsic or acquired. Intrinsic resistance is a naturally occurring trait and can be considered as a species characteristic, whereas acquired resistance derives either from genetic mutations or recombinations. Lactobacilli can display a wide range of intrinsic antibiotic resistances [9], but in most cases these resistances are not transferable. *Lactobacillus* strains with non-transferable antibiotic resistances usually do not represent a safety concern. To some extent, antibiotic resistance might even be a useful property if the probiotic strain is to be used as a prophylactic agent in the treatment of antibiotic associated diarrhea [6]. Although plasmid and transposon-linked antibiotic resistances are not very common among lactobacilli, they do occur [16] and their safety implications should be taken into consideration. Since transfer of antibiotic resistance genes may occur between phylogenetically distant bacteria [8], strains harboring mobile elements carrying resistance genes should be carefully assessed if they are to be used as human or animal probiotics. Finally, as obliged for pharmaceutical products, human studies assessing possible side-effects should also be performed for other types of probiotic products, as well as post-market epidemiological surveillance of possible adverse incidents.

3.3. Functional aspects of probiotic strains

From the producer's perspective, the potential health benefits of new probiotic strains are of utmost importance, and the majority of research is invested in the substantiation of health claims. However, also a number of more fundamental aspects determine the functional qualities of a probiotic strain including resistance to gastric acidity and bile toxicity, adhesion to gut epithelial tissue, ability to colonize the GI-tract and ability to modulate immune responses.

Before reaching the intestinal tract, probiotic bacteria must first pass the stomach barrier where the secretion of gastric acid constitutes a primary defense mechanism against most ingested microorganisms. In vitro tests usually include the assessment of bacterial growth on deMan, Rogossa and Sharpe (MRS) medium amended with pepsin and hydrogen chloride to pH values of between 2.0 and 3.4. However, survival of bacterial strains in human gastric juice is a more accurate indication of the ability of strains to survive passage through the stomach, for which bifidobacteria proved less successful than lactobacilli [12].

Besides gastric juice, probiotic strains should also possess a certain resistance against the effects of bile acids encountered in the small intestine. Bile acids are synthesized in the liver from cholesterol and are secreted from the gall bladder into the duodenum in a conjugated form (500-700 ml/day) [21]. These acids then undergo extensive chemical modifications in the colon (deconjugation, dehydroxylation, dehydrogenation and deglucuronidation), almost solely as a result of microbial activity [24]. Both conjugated and deconjugated bile acids exhibit antibacterial activity, however, the deconjugated forms are more inhibitory. Gram-positive bacteria are found to be more sensitive than Gram-negative bacteria [14]. Bile acid resistance, based on relative ability to grow in the presence of bile acids, varies greatly among strains of each species of probiotic bacteria [17]. Therefore, an extensive selection of the most resistant strain(s) is necessary.

Adhesion of probiotic bacteria to the intestinal surface and the subsequent colonization of the human GI-tract have been suggested as important prerequisites for probiotic action. Adherent strains are likely to persist longer in the intestinal tract and thus have better possibilities to exert metabolic and immunomodulatory effects compared to non-adhering strains [37]. Adhesion may also provide means of competitive exclusion of pathogenic bacteria from the intestinal epithelium. HT-29 and Caco-2 cells are human cell lines expressing morphological and physiological characteristics of normal human enterocytes that have been exploited to elucidate the mechanisms mediating enteropathogen adhesion [7]. More recently, these cell lines have been used to select for and subsequently assess lactic acid bacteria on the basis of their adhesion properties [46]. Despite the availability of various probiotic adhesion studies, the role of adhesion in successful probiotic functioning remains speculative [36]. It could also be argued that strong adhesion ability may increase the risk of infection in the host [4]. Some probiotic strains are poorly adhering in vitro and/or in vivo but still show positive effects on the host. Regardless of the adhesion capacity of a probiotic strain, it is now widely accepted that administration of a daily dose of probiotics is the most efficient way to maintain probiotic activity in the gut. However, discussions on the quantity of this daily dose are still ongoing.

Besides the criteria related to the potential of a strain to reach the GI-tract in a metabolically active state, the most crucial and certainly the most difficult research area is to determine the potential health benefits of a probiotic strain. In general, two subsequent research tasks need to be performed in order to document probiotic activity. Most crucial is the setup of clinical trials analyzing the beneficial effects of a probiotic to a group of test persons. Secondly, if a probiotic effect has been reported, the microbial, biochemical and molecular mechanisms responsible for the observed effect should be unraveled. The levels of evidence for health benefits depend on the study design and the methodological quality. The most objective and thus reliable evidence can be obtained from randomized, double-blind, placebo-controlled studies monitoring a sufficiently large test group of subjects that is placed under dietary supervision [27]. A general recommendation for the testing of probiotics is that the placebo consists of the food carrier without the test probiotic. The principle outcome of efficacy studies on probiotics should be proven benefits such as (i) statistically and biologically significant improvement in physical condition, symptoms, signs, well-being or quality of life; (ii) reduced risk of disease or longer time to next occurrence, or (iii) faster recovery from illness. Each of these symptoms should have a proven correlation with the probiotic tested [13]. Furthermore, it is also desirable that these studies are performed by independent scientific institutions and that results are published in peer-reviewed journals addressing the appropriate research fields [32], including publications on negative results.

Numerous (potential) probiotic effects have been reported [31, 32], some better documented than others often related to the complexity of the research. The strength of evidence for positive effects of probiotics in intestinal disorders appears to be most convincing in cases of antibiotic-associated diarrhea, gastroenteritis and lactose intolerance. Evidence is increasing, although more slowly, for inflammatory bowel disease (IBD) and intestinal infections [27]. However, one strain should not be expected to produce all potential benefits. Clearly, each specific disorder will require a careful selection of the most suitable probiotic strain in order to achieve the optimal health benefits [17]. As a result of these strain-dependent health benefits, studies are frequently performed using a mixture of multiple probiotic strains, for example in the prevention of chronic pouchitis [18].

4. Exploitation of new probiotic strains

To succeed in promoting the consumption of probiotic products, the food industry has to meet the demands of the consumer that all probiotic foods should be safe, functional and possess attractive sensory properties. Before probiotic strains can be delivered to consumers, they must first be tested for their ability to be included in the product matrix of choice. Subsequently, the introduced strains need to survive and retain their functionality during storage as frozen or freeze dried cultures as well as in the food products into which they are commercialized. Finally, the packaging material and storage conditions are important factors influencing the quality of the probiotic product during its shelf life. As a result of their long history, fermented dairy products are by far the most popular formulation of probiotic foods [42]. However, the use of modern technologies has also resulted in other product types such as powders, capsules and tablets, which favor oral administration in situations where dairy based products can not be applied. Commercially available probiotic cultures may consist of a single strain or a mixture of several strains. In most cases, the probiotic properties are affected by the way in which the strain or culture has been produced [15], meaning that each strain should be characterized extensively in order to allow an effective optimization of the production process.

In case of fermented dairy products, reliable acid-forming ability is the most important characteristic in selecting starter microorganisms. When selecting probiotics, however, selection criteria should be linked to the impact on human health and well-being. Because the environment within the mammalian GI-tract might be quite different from the food matrix environment, the probiotic is often not suitable as a starter organism [15]. Therefore, it is quite common to use probiotic bacteria mixed together with other bacteria, e.g. *Lactobacillus delbrueckii* subsp. *bulgaricus* and *Streptococcus thermophilus*, in order to obtain the desired flavor and texture. Application of probiotic cultures in non-dairy products such as beer, fruit juices or cake fillings represents an additional challenge in the field of viability and storage [3]. Because probiotic cultures are present as additives in these kinds of products and usually do not grow out to larger numbers, this sets specific demands for the probiotic stability. Physico-chemical factors such as water activity, oxygen tension and temperature become increasingly important when dealing with these kinds of products. For instance, storage at room temperature of freeze-dried probiotic products is made possible by using advanced encapsulation technologies to ensure viability and stability of the probiotic cultures [30]. More detailed information on different aspects of probiotic production technology can be found in the review of Saarela and co-workers [36].

Next to the search for functional and safe probiotic strains, it is equally important to realize that also the product containing the probiotic organism(s) needs to live up to a number of minimal standards. Besides verification of possible side-effects and effectiveness related to the use of different types of product, a profound quality control and labeling legislation needs to be established for commercial probiotic products worldwide. In most countries, only general health claims are allowed on foods containing probiotics. An FAO/WHO Working Group recently recommended that specific health claims relating to probiotic effects should be allowed when sufficient scientific evidence is available [13]. The Working Group recommends that it should be the responsibility of the manufacturer to establish health claims as truthful and not misleading, by means of an independent third party review composed of scientific experts in the field. From the consumer's point of view, the product label should contain information on genus, species and strain designation; minimum viable numbers of each probiotic strain at the end of shelf-life; effective dose; health claims; storage conditions and corporate contact details.

Besides legislation of health claims and labeling issues, similar regulations need to be established regarding quality control assessment during and after the production process. The acronym "GMP" (Good Manufacturing Practice) is used internationally to describe a set of principles and procedures, which, when followed by manufacturers of mainly therapeutic goods, helps ensure that the products manufactured, will have the required quality. There is a tending to extrapolate this GMP concept towards the production of probiotic products, although no profound guidelines have been set up yet and quality control is the sole responsibility of the manufacturer. In general, a comprehensive approach to determine shelf life of probiotic bacteria is linked to maintenance of efficacy. In the assumption that efficacy is linked to viability, literature data suggest that the quality of probiotic-containing products with regard to microbial content and shelf life can be highly variable [39, 43, 44]. The need for a sound legislation setting guidelines for quality control of probiotic manufacturing is intertwined with the development, validation and harmonization of reliable analysis methods in order to allow comparison of quality scores on an international level.

5. Conclusion

The growth and expansion of the probiotic market is not expected to come to a halt in the near future due to the increasing list of health claims arising from scientific research. Given the relatively limited diversity of probiotic strains currently exploited, the search for new probiotic microorganisms and the subsequent development of new products with specific health benefits is one of the major aims of the functional food industry. Hopefully, manufacturers will realize that both organism and product need to live up to a large number of criteria before they can reinforce the probiotic market on a reliable and effective basis.

References

1. Adams, M.R. and Marteau, P. (1995). On the safety of lactic acid bacteria. *International Journal of Food Microbiology,* **27**: 263-264.
2. Adams, M.R. (1999). Safety of industrial lactic acid bacteria. *Journal of Biotechnology,* **68**: 171-178.
3. Andersen, L. (1998). Pre- and probiotics – sausage science. *Functional Foods,* June: 26-29.
4. Apostolou, E., Kirjavainen, P.V., Saxelin, M., Rautelin, H., Valtonen, V., Salminen, S. and Ouwehand, A.C. (2001). Good adhesion properties of probiotics: a potential risk for bacteremia? *FEMS Immunology and Medical Microbiology,* **31**: 35-39.
5. Axelsson, L. (1998). Lactic acid bacteria: classification and physiologie. In *Lactic acid bacteria, microbiology and functional aspects*, 2nd edition, (Salminen, S. and von Wright, A., Eds.). Marcel Dekker, Inc., New York, USA. pp.1-59.
6. Charteris, W.P., Kelly, P.M., Morelli, L. and Collins, J.K. (1998). Antibiotic Susceptibility of Potentially Probiotic *Lactobacillus* Species. *Journal of Food Protection,* **61**:1636-1643.
7. Cocconier, M.H., Bernet, M.F., Kerneis, S., Chauviere, G., Fourniat, J. and Servin, A.L. (1993). Inhibition of adhesion of enteroinvasive pathogens to human intestinal Caco-2 cells by *Lactobacillus acidophilus* strain LB decreases bacterial invasion. *FEMS Microbiology Letters,* **110**:299-305.
8. Courvalin, P. (1994). Transfer of antibiotic resistance genes between gram-positive and gram-negative bacteria. *Antimicrobial Agents and Chemotherapy,* **38**:1447-1451.
9. Danielsen, M. and Wind, A. (2003). Susceptibility of *Lactobacillus* spp. to antimicrobial agents. *International Journal of Food Microbiology,* **82**:1:1-11.
10. Döderlein, A. (1892). Das Scheidensekret und seine Bedeutung für das Puerperalfieber. (The vaginal transsudate and its significance for childbed fever.) *Centralblatt für Bacteriologie,* **11**:699-700.
11. Donohue, D.C. and Salminen, S.J. (1996). Safety of probiotic bacteria. *Asian Pacific Journal of Clinical Nutrition,* **5**:25-28.
12. Dunne, C., O'Mahony, L., Murphy, L., Thornton, G., Morrissey, D., O'Halloran, S., Feeney, M., Flynn, S., Fitzgerald, G., Daly, C., Kiely, B., O'Sullivan, G.C., Shanahan, F. and Collins, J.K. (2001). *In Vitro* Selection Criteria for Probiotic Bacteria of Human Origin: Correlation With *in Vivo* Findings. *American Journal of Clinical Nutrition,* **73**(S): 386S-392S.
13. FAO/WHO Working Group Report on Drafting Guidelines for the Evaluation of Probiotics in Food. London Ontario, Canada, April 30 - May 1, 2002.
14. Floch, M.H., Binder, H.J., Filburn, B. and Gershengoren, W. (1972). The effect of bile acids on intestinal microflora. *American Journal of Clinical Nutrition,* **25**:1418-1426.
15. German, B., Schiffrin, E.J., Reniero, R., Mollet, B., Pfeifer, A. and Neeser, J.-R. (1999). The development of functional foods: lessons from the gut. *TIBTECH,* **17**:492-499.
16. Gevers, D., Huys, G., Swings, J. (2003). *In vitro* conjugal transfer of tetracycline resistance from *Lactobacillus* isolates to other Gram-positive bacteria. *FEMS Microbiology Letters,* **225** (1): 125-130.

17. Gilliland, S. (2002).Technological and commercial applications of lactic acid bacteria; health and nutritional benefits in dairy products. http://www.fao.org/es/ESN/food/foodandfood_probio_en.stm

18. Gionchetti, P., Rizzello, F., Venturi, A., Brigidi, P., Matteuzzi, D., Bazzocchi, G., Poggioli, G., Miglioli, M. and Campieri, M. (2000). Oral bacteriotherapy as maintenance treatment in patients with chronic pouchitis: A double-blind, placebo-controlled trial. *Gastroenterology,* **119**: 2: 305-309.

19. Goldin, B.R. and Gorbach, S.L. (1992). Probiotics for human consumption. In *Probiotics, The Scientific Basis (*Fuller R., Ed.). London: Chapman and Hall pp.355-376.

20. Grand, M., Kuffer, M. and Baumgartner, A. (2003). Quantitative analysis and molecular identification of bifidobacteria strains in probiotic milk products. *European Food Research and Technology,* **217**:1: 90-92.

21. Hoffman, A.F., Molino, G., Milanese, M. and Belforte, G. (1983). Description and stimulation of a physiological pharmokinetic model for the metabolism and enterohepatic circulation of bile acids in man. *Journal of Clinical Investigation,* **71**:1003-1022.

22. Holzapfel, W.H., Haberer, P., Snel, J., Schillinger, U. and Huis in't Veld, J.H.J. (1998). Overview of gut flora and probiotics. *International Journal of Food Microbiology,* **41** :85-101.

23. Holzapfel, W.H., Haberer, P., Geisen, R., Bjorkroth, J. and Schillinger, U. (2001). Taxonomy and important features of probiotic microorganisms in food and nutrition. *American Journal of Clinical Nutrition,* **73**:2: 365S-373S.

24. Hylemon, P.B. and Glass, T.L. (1983). Biotransformation of bile acids and cholesterol by the intestinal microflora. In *Human intestinal microflora in health and disease,* (Hentges, D.J., Ed). New York: Academic Press, pp.189-213.

25. Lilley, D.M. and Stillwell, R.H. (1965). Probiotics: growth promoting factors produced by microorganisms, *Science,* **147**:747-748.

26. Lucas, J. (2002). EU-funded research on functional foods. *British Journal of Nutrition,* **88**(S): S131-S132.

27. Marteau, P., Seksik, P. and Jian, R. (2002). Probiotics and intestinal health effects: a clinical perspective. *British Journal of Nutrition,* **88**(S1):S51-S57.

28. McFarland, L. V., and Elmer, G. W. (1995). Biotherapeutic agents: past, present and future. *Microecology and Therapy,* **23**: 46–73.

29. Metchnikoff, E. (1907). The prolongation of life, Heinemann, London.

30. Myllärinen, P., Forssell, P., von Wright, A., Alander, M. and Matilla-Sandholm, T. (1998). The use of starch as a capsulation material for lactic acid bacteria. In Matilla-Sandholm, T. and Kauppila, T. (Eds.) *Functional Food Research in Europe*, 3rd workshop, FAIR CT96-1028. PROB-DEMO, VTT Symposium 187, Haikko, Finland, p.91.

31. Naidu, A.S., Bidlack, W.R. and Clemens, R.A. (1999). Probiotic spectra of lactic acid bacteria (LAB). *Critical Reviews of Food Science and Nutrition,* **39**:13-126.

32. Ouwenhand, A.C., Salminen, S. and Isolauri, E. (2002). Probiotics: an overview of beneficial effects. *Antonie van Leeuwenhoek,* **82**: 279-289.

33. Reid, G. (2002). FAO/WHO Expert Consultation on Evaluation of Health and Nutritional Properties of Powder Milk with Live Lactic acid bacteria. Regulatory and clinical aspects of dairy probiotics. Lawson Health Research Institute, London, Ontario, Canada.
34. Reid, G., Jass, J., Sebulsky, M. and McCormick, J. (2003). Potential Uses of Probiotics in Clinical Practice. *Clinical Microbiology Reviews,* **16**:4: 658-672.
35. Reuter, G. (2001). Probiotics - Possibilities and limitations of their application in food, animal feed, and in pharmaceutical preparations for man and animals. *Berliner und Munchener Tierarztliche Wochenschrift,* **114**:11-12: 410-419.
36. Saarela, M., Mogensen, G., Fondén, R., Mättö, J. and Matilla-Sandholm, T. (2000). Probiotic bacteria: safety, functional and technological properties. *Journal of Biotechnology,* **84**: 197-215.
37. Salminen, S., Isolauri, E. and Salminen, E. (1996). Clinical uses of probiotics for stabilizing the gut mucosal barrier: Succesfull strains for future challenges. *Antonie Van Leeuwenhoek,* **70**: 347-358.
38. Salminen, S., Von Wright, A., Morell, L., Marteau, P., Brassart, D., De Vos, W. M., Fonden, R., Saxelin, M., Collins, K., Mogensen, G., Birkeland, S.E. and Mattila-Sandholm, T. (1998a). Demonstration of Safety of Probiotics - a Review. *International Journal of Food Microbiology,* **44**: 93-106.
39. Sanders, M.E. and Huis in't Veld, J. (1999). Bringing a probiotic-containing functional food to the market: microbiological, product, regulatory and labeling issues. Antonie Van Leeuwenhoek. *International Journal of General and Molecular Microbiology,* **76**: 293-315.
40. SCAN (2000): Report of the Scientific Committee on Animal Nutrition on the Safety of Use of Bacillus Species in Animal Nutrition. European Commission Health & Consumer Protection Directorate General. http://europa.eu.int/comm/food/fs/sc/scan/out41.pdf
41. Stanton, C., Gardiner, G., Meehan, H., Collins, K., Fitzgerald, G., Lynch, P. B. and Ross, R.P (2001). Market Potential for Probiotics. *American Journal of Clinical Nutrition,* **73**: 476S-483S.
42. Svensson, U. (1999). Industrial perspectives. In *Probiotics: A critical review*., (Tannock G.W., Ed.) Horizon Scientific Press. Wymondham, pp.57-64.
43. Temmerman, R., Scheirlinck, I., Huys, G. and Swings, J. (2003). Culture-independent Analysis of Probiotic Products using Denaturing Gradient Gel Electrophoresis (DGGE). *Applied and Environmental Microbiology,* **69**:1: 220-226.
44. Temmerman, R., Huys, G. and Swings, J. (2004). Identification of Lactic Acid Bacteria: Culture-dependent and Culture-independent Methods. *Trends in Food Science and Technology,* **15**: 348-359.
45. Threlfall, E., Ward, L, Frost, J. and Willshaw, G. (2000). The emergence and spread of antibiotic resistance in food-borne bacteria. *International Journal of Food Microbiology,* **62**:1-5.
46. Tuomola, E.M. and Salminen, S. (1998). Adhesion of some probiotic and dairy *Lactobacillus* strains to Caco-2 cell cultures. *International Journal of Food Microbiology,* **41**: 45-51.
47. Verschuere, L., Rombaut, G., Sorgeloos, P. and Verstraete, W. (2000). Probiotic bacteria as biological control agents in aquaculture. *Microbiology and Molecular Biology Reviews,* **64**(4): 651-671.

Taxonomy and Culture Collections

Uniqueness of the "Smart State's" microbial diversity

From a Microbial Library to Biodiscovery An academic initiative complementing Queensland Government's Action Plan

D.I. Kurtböke

Institute for Sustainability, Health and Regional Engagement,
University of the Sunshine Coast, Maroochydore DC, Qld 4558, Australia

1. Introduction

Australia is one of the world's most biodiverse continents. It has been geologically separated from other continents for over 20 million years, which has allowed a period of extensive evolutionary divergence. As a result, Australia has a very high rate of endemism in its flora, fauna and associated microflora. In addition, the Australian continent contains diverse range of biogeographical regions including an arid interior, tropical monsoon areas in the north, Mediterranean and temperate climates in the south, alpine regions in Tasmania and in the Southeast. Australian marine environments, which are exposed to 3 different major ocean currents, appear to be the most diverse in the world.

As a consequence of the realization of the importance of unique Australian flora and fauna for potential pharmaceutical applications, the Australian Federal and State Governments are now taking steps towards the commercialization of this diversity. The sections below (2, 3, and 3.1) outline the policies of the Queensland Government regarding biodiscovery.

2. Biotechnology and Bioindustries in the "Smart State"

Queensland's strength in biotechnology comes from its unusual biodiversity, which includes 19 terrestrial bioregions and 17 marine bioregions, as well as five world heritage regions covering 40 million hectares (http://www.sdi.qld.gov.au/innovation/biotechnology/default.asp). Some 80% of Queensland's marine and terrestrial species are found nowhere else in the world and almost 4% of the world's biodiversity is unique to Queensland. As a result, Queensland's biodiversity has led to an accumulation of well-developed skills and know-how in the management of ecologically-sensitive regions as well as in the regulated exploitation of its flora and fauna for medical, pharmaceutical and agricultural research. The combination of natural resources and a powerful infrastructure sets Queensland apart and attracts overseas collaboration. In order to accelerate the development of sustainable bioindustries, the Queensland Government announced a $270 million 10-year **Bioindustries Strategy** in May 1999. This funding is being used by **the Department of State Development and Innovation** to position the State as a centre of biotechnology in the Asia-Pacific region (http://www.sdi.qld.gov.au/innovation/publications/smartstate/default.asp.; http://www.sdi.qld.gov.au/innovation/biotechnology/bioindustries.asp).

In order to assist in its implementation, the Government has also established a **Biotechnology Advisory Council** bringing together representatives from research and academic institutions, government and the private sector to provide independent strategic advice to the Queensland Government on the State's burgeoning biotechnology industry [1]. In addition, the Queensland Government has entered into **the Australian Biotechnology Alliance** for ten years with the two other biotechnologically advanced states in Australia, namely Victoria and New South Wales, which are home to some of the largest biotech research centres in the Southern Hemisphere. This alliance formalises existing links and collaborative efforts between the States and builds on a highly skilled workforce, internationally recognized scientists and research facilities, stable structures, proximity to Asia-Pacific markets and low research and business costs (http://www.sdi.qld.gov.au/innovation/biotechnology/bioindustries.asp).

Furthermore, the New Zealand-Queensland Biotechnology Collaboration agreement was signed to facilitate information exchange between research institutes, universities and businesses allowing the New Zealand and Queensland biotechnology sectors to secure joint commercial advantage from closer collaboration [1].

3. Biodiscovery in the "Smart State"

In line with the **United Nations Environment Program Convention on Biological Diversity** (CBD), to which Australia is signatory, the Queensland Government has introduced a **Biodiscovery Bill**, an Australian first, to regulate the collection of samples associated with the reefs, rainforests, savannas and wetlands of the State while providing the basis for a biodiscovery industry that is already developing new bio-products such as medicines and natural pesticides [6]. The Queensland Government is also developing legislation to establish a more efficient permitting regime for the use of the State's native biological resources. The aim of this regime is a) to establish a **Biodiscovery Collection Authority** to govern the sustainable collection of native biological resources from the State's land and water; b) to introduce **Benefit Sharing Agreements** to give Queenslanders a fair share of the benefits derived from the State's natural resources (http://www.sdi.qld.gov.au/innovation/biotechnology/bioindustries.asp).

The Biodiscovery Fund and **Smart State Research Facility Fund** are other initiatives aimed at stimulating biodiscovery research in Queensland including the establishment of an Australian Tropical Forest Institute in Cairns. In addition, the Queensland Government provides support to the research centres in the State including Cooperative Research Centres in which most of the biodiscovery is carried out. As a consequence of such support and funding from the Queensland Government, it is expected that the State's microbiota will receive increased attention in the next decade on the national, regional and international levels (http://www.sdi.qld.gov.au/innovation/biotechnology/bioindustries.asp).

3.1. The Smart State's unexplored microbial diversity

Although its geographical isolation has resulted in the evolution of a different microflora in Australia including many novel bacteria and fungi, still very little is known about the endemism of the Australian microbial, in particular, Actinomycete communities. It has been estimated that so far $<1\%$ of all microorganisms existing has been identified and characterized [2]. Taxonomic relationships between microbes are still lacking. Consequently, the need for the establishment of a **Microbial Genetic Resources Research Network** to accelerate research on microorganisms of medicinal, agricultural and industrial importance has been emphasized [5], and the groundwork is underway (see Sly, Chapter 6). The resulting network of scientists and the scope of the operation can be viewed at http://www.cbit.uq.edu.au/amrrn/scope.htm.

Studies mapping the State's microbial diversity reveal substrates that contain rare and novel microorganisms [2,3,4]. There is, therefore, an exciting opportunity within Queensland to make a significant contribution to the field of drug discovery, agriculture, and food and in the other areas of biotechnology through the acquisition and screening of this microbiota.

Examples include the detection of novel microorganisms producing novel compounds in the naturally occurring suppressive soils and mulches of the Tropical Rain Forest of Queensland and on the phyllosphere of the native trees [2,3,4]. *Saccharomonospora* spp. rich in enzyme production were isolated from Queensland's sugar cane bagasse [3]. Endophytic microorganisms associated with the tropical rain forest and subtropical plants yielded *Microbispora, Microtetraspora, Planomonospora* and *Planobispora* spp. as well as the antimicrobial compound producing *Micromonospora* spp. [2,3,4]. The detection of these novel taxa in the Smart State environments provides evidence for the occurrence of bioactive compound producing novel microorganisms in Queensland.

4. Creation of a Microbial Library

In order to complement the Smart State's initiatives, a microbial library has been established at the University of the Sunshine Coast utilizing the geographical resources of the most biodiverse State of Australia. The library contains mostly actinomycetes, non-actinomycete bacteria, fungi and bacteriophages and develops as part of the large Microbial Genetic Resources Network, which is now being established in the State with the participation of the State's other expert Institutes (http://www.cbit.uq.edu.au/amrrn/scope.htm) (see Sly, Chapter 6).

4.1. Projects contributing to the Microbial Library

4.1.1. Pharmaceuticals

The Microbial Library has now, through the use of effective conventional and molecular detection procedures and focusing principally on the State's unique habitats and microenvironments, detected rare and novel actinomycetes as potential sources of novel therapeutic agents.

The collection now holds over 3000 actinomycetes representing over 75% of novelty (Figs 1 and 2) and showing strong antimicrobial activity in particular against vancomycin and methycillin resistant bacteria originating from guts of the unique Australian termite families, phyllosphere and rhizosphere of sub-tropical native plants, aerial roots of the subtropical trees, marine sediments and marine macroorganisms (Table 1).

Compounds are also being studied towards other targets such as immunosuppressants and anti-cancer compounds as well as enzymes and agrobiologicals as part of the international agreements the Microbial Library has attracted.

Figure 1: Phylogenetic analysis of the USC's isolate from *Coptotermite* species gut

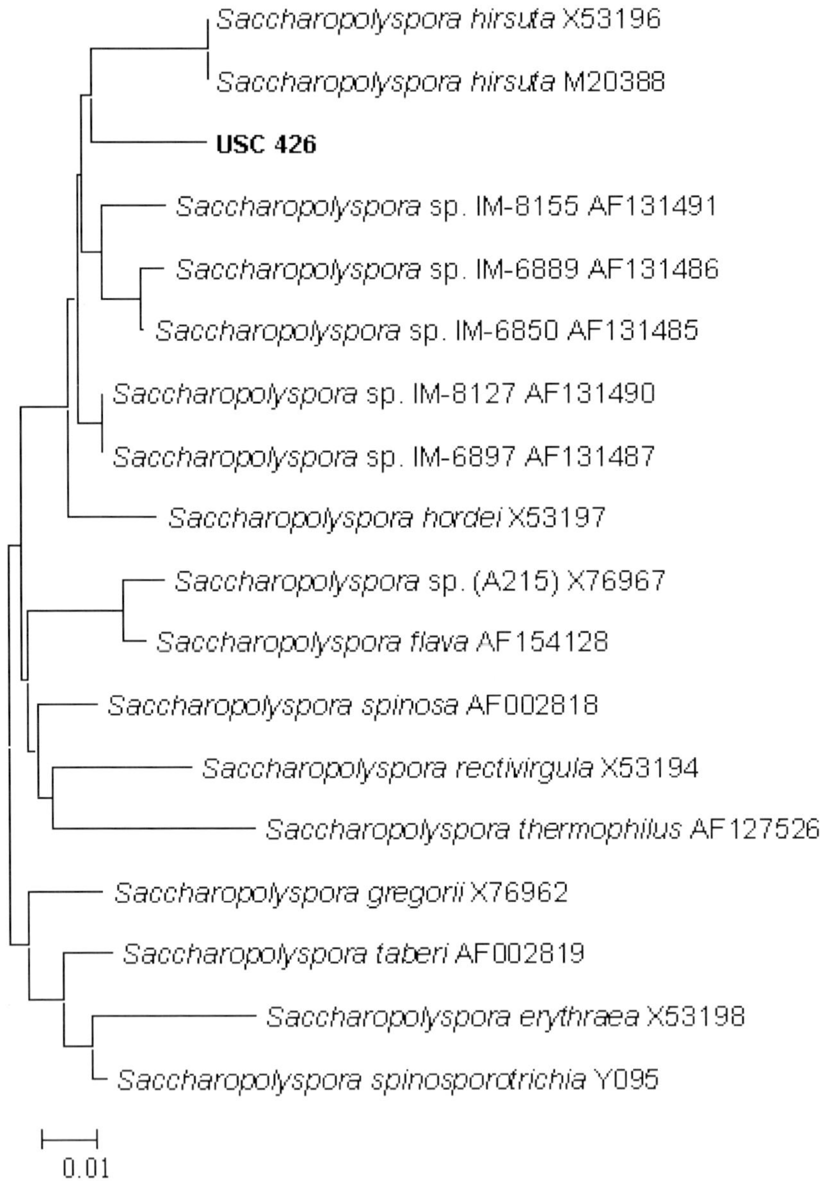

Figure 2: Phylogenetic analysis of the USC's isolate from *Nasutotermite* species gut

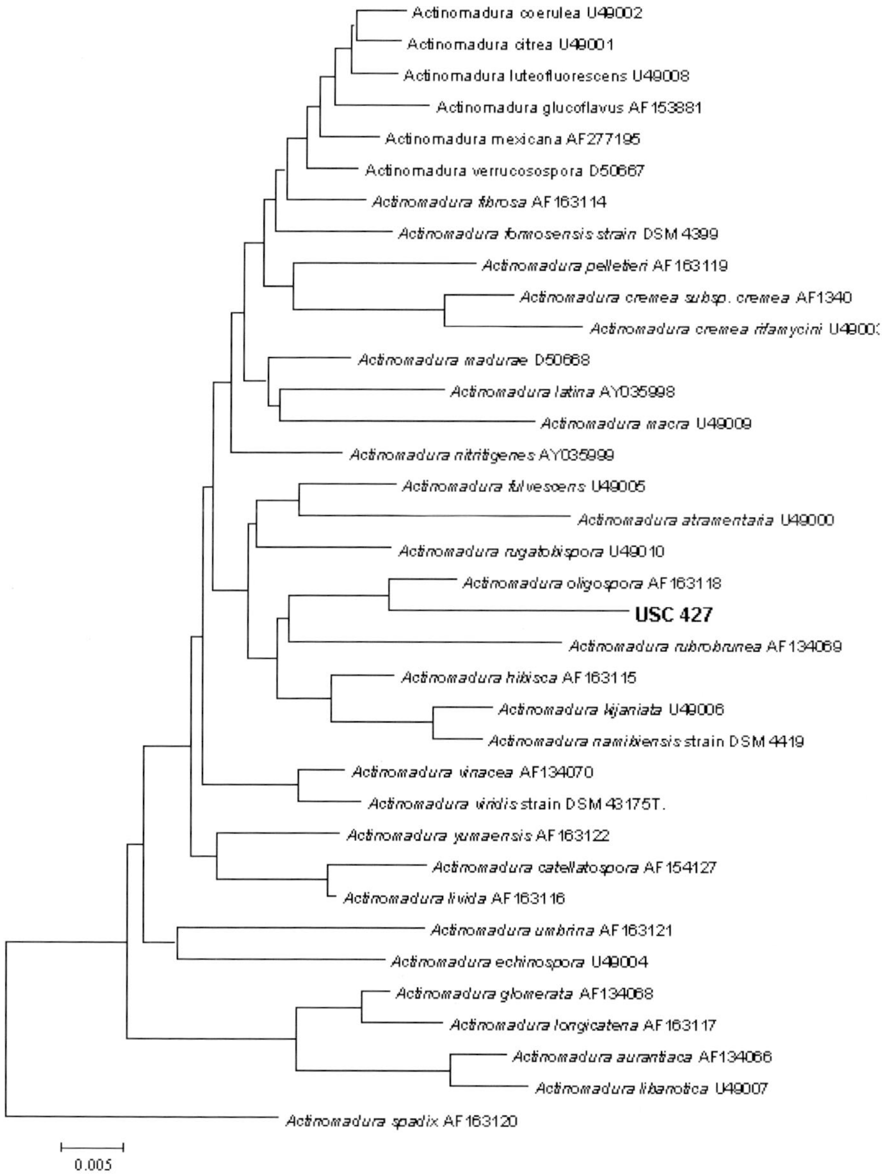

Table 1: Types of habitats searched for bioactive compound producing actinomycetes in Queensland

Habitat type	Habitat code
Agricultural soils	HB-AGS
Soils	HB-S
Organic mulches	HB-OM
Bagasse	HB-B
Termite guts	HB-TG
Termite mounds	HB-TM
Native plant phyllosphere	HB-NPP
Native plant rhizosphere	HB-NPR
Marine sediments	HB-MS
River sediments	HB-RS
Aerial roots of subtropical trees	HB-ARST
Composted poultry manure	HB-PM
Composted pig manure	HB-PGM
Mangrove muds	HB-MM
Tree barks	HB-TB

4.1.2. Agrobiologicals and Biological Control Agents

Many fungal pathogens cause significant destruction to crops and native plants in Australia resulting in significant losses for Australian Farmers. In order to control these pathogens different control measures are employed in Australia and biological control is one of them [2]. However, while the eradication of a pathogen by a biological control agent may be the ideal solution for disease control, even a balanced equilibrium between a pathogen and its antagonist(s) should be considered beneficial. As a result, antifungal actinomycetes are continuously being screened against the destructive fungal pathogens infecting Queensland's crops.

Furthermore, emerging ecosystem and public health issues in Queensland such as the need to replace toxic agrochemicals with pest-specific agrobiological compounds provides support towards the search for biopesticide-producer microorganisms. Accordingly, actinomycetes with antagonistic activity towards common Australian plant pathogens are being isolated and tested for agrobiological activity to successfully control plant pathogens. There is now evidence to suggest that diverse ranges of soil environments in Queensland yield different microbial flora including rare actinomycetes antagonistic towards plant pathogenic fungi (Table 3). In addition, other substrates such as plant roots, phyllospheres and carpospheres harbour these antagonists. Zoospore forming actinomycetes such as *Actinoplanes* spp. isolated from the rhizoplane were found to be more effective in controlling the growth of oospore forming fungi such as *Pythium and Phytophthora* spp. Hyperparasitism was also common amongst the carposphere and phyllosphere actinomycetes such as coiling around *Botyritis cinerea* hyphae. Whereas, antibiosis was recorded from actinomycetes isolated from the pathogen infested soils, for example, against *Rhizoctonia solani and Geaumannomyces graminis var. tritici,* (Table 3).

Since Queensland actinomycetes offer significant potential for biological control applications, further mapping of this unique microflora will encourage the development of strategies in the State to employ effective biological control systems.

4.1.3. Bioremediation, Biodegradation and Bioconversion

The Microbial Library contains microorganisms in particular actinomycetes such as *Lechevalieria, Nonomurae*, *Thermomonospora*, *Streptomyces* and *Actinomadura*, active in composting and wood decaying through the production of a series of non-specific extracellular peroxidases that are involved not only in ligninolysis, but also attack a broad range of aromatic and phenolic compounds.

The focus of the Microbial Library also extends to the discovery of novel and stable enzymes from Australian biological resources to the breakdown of urban and agricultural waste, bioremediate tannery and paper industry waste, as well as the use of these enzymes in detergents, food and textile manufacturing.

Thermomonospora and thermophilic *Nocardioides* spp. were isolated from the composted waste of a local tannery in the Sunshine Coast Region. These microorganisms show a wide range of enzymatic activity for application in the bioremediation of recalcitrant xenobiotic compounds such as polyaromatic and polychlorinated aromatic hydrocarbons.

Table 2: Examples of the selected actinomycete taxa from the USC's Microbial Library showing antimicrobial activity

SAMPLE TYPE															
GENERA	**HB-AGS**	**HB-S**	**HB-OM**	**HB-B**	**HB-TG**	**HB-TM**	**HB-NPP**	**HB-NPR**	**HB-MS**	**HB-RS**	**HB-ARST**	**HB-PM**	**HB-PGM**	**HB-MM**	**HB-TB**
Streptomyces	+	+	+	+	+	+	+	+	+	+	+	+	+	+	+
Actinobifida	-	-	+	+	-	-	-	-	-	-	-	+	+	-	+
Actinocorallia	+	+	-	-	-	-	-	-	-	-	-	-	-	+	-
Actinobispora	+	+	-	-	-	-	-	-	-	-	-	-	-	+	-
Actinomadura	+	+	+	+	+	+	-	+	-	-	-	+	+	+	-
Actinoplanes	+	+	-	-	+	+	+	+	+	+	-	-	-	+	-
Amycolatopsis	+	+	+	-	-	+	-	+	-	-	-	-	-	+	-
Catellatospora	+	+	-	-	-	-	-	+	-	-	-	-	-	+	-
Catenuloplanes	+	-	-	-	-	-	-	-	-	-	-	-	-	-	-
Couchioplanes	-	-	-	-	-	-	-	-	+	+	-	-	-	+	-
Dactylosporangium	+	+	+	-	+	+	-	+	-	-	-	-	-	+	-
Clavibacter	-	-	-	-	-	-	-	+	+	-	-	-	-	-	-
Curtobacterium	-	-	-	-	-	-	-	+	+	-	-	-	-	-	-
Dietzia	+	+	-	-	-	+	-	-	+	+	+	+	+	+	+
Excellospora	+	+	+	+	+	+	-	+	-	-	-	+	+	+	+
Geodermatophilus	+	+	-	-	+	+	+	+	-	-	-	-	-	+	-
Kibdelosporangium	+	+	+	-	+	+	-	+	-	-	-	-	-	+	-
Kitasatospora	+	+	+	-	+	+	-	+	-	-	-	-	-	+	-
Kocuria	+	+	+	-	-	-	-	+	-	-	-	-	-	+	-
Kutzneria	+	+	+	-	-	+	-	+	-	-	-	-	-	+	-
Kytococcus	-	-	-	-	-	-	-	-	+	+	-	-	-	+	-
Lechealiera	-	-	+	+	-	-	-	-	-	-	-	-	-	-	+

Table 2 continued

SAMPLE TYPE															
GENERA	**HB-AGS**	**HB-S**	**HB-OM**	**HB-B**	**HB-TG**	**HB-TM**	**HB-NPP**	**HB-NPR**	**HB-MS**	**HB-RS**	**HB-ARST**	**HB-PM**	**HB-PGM**	**HB-MM**	**HB-TB**
Lentzea	+	+	-	-	+	+	+	+	-	-	-	-	-	+	-
Microbispora	+	+	-	-	+	+	+	+	-	-	-	-	-	+	-
Micromonospora	+	+	+	+	+	+	+	+	+	+	+	+	+	+	+
Microtetraspora	+	+	-	-	+	+	+	+	-	-	-	-	-	+	-
Nesterenkonia	-	-	-	-	-	-	-	-	+	+	-	-	-	+	-
Nocardia	+	+	+	-	+	+	+	+	+	+	+	+	+	+	+
Nocardiopsis	-	-	+	+	-	-	-	-	-	-	-	+	+	-	+
Nocardioides	+	+	+	+	+	+	+	+	+	+	+	+	+	+	+
Nonomuraea	+	+	+	-	+	+	-	+	-	-	-	+	+	+	+
Pilimelia	+	+	-	-	+	+	+	+	-	-	-	-	-	+	-
Planomonospora	+	+	+	-	-	+	-	+	-	-	+	-	-	+	-
Planotetraspora	+	+	+	-	-	+	-	+	-	-	+	-	-	+	-
Prauserella	+	+	+	-	-	+	-	+	-	-	-	-	-	+	-
Rhodococcus	+	+	+	-	+	+	+	+	+	+	+	+	+	+	+
Saccharomonospora	-	-	+	+	-	-	-	-	-	-	-	+	+	-	+
Saccharopolyspora	+	+	+	-	+	+	-	-	-	-	-	+	+	+	+
Streptomyces	+	+	+	+	+	+	+	+	+	+	+	+	+	+	+
Streptosporangium	+	+	+	+	+	+	+	-	-	-	-	-	+	+	+
Terracoccus	+	+	+	-	+	+	-	+	-	-	-	-	-	+	-
Verrucosispora	-	-	-	-	-	-	-	-	-	-	-	-	+	-	-
Williamsia	-	-	-	-	-	-	-	-	+	+	-	-	-	+	-

+: Detected in that habitat
- : not detected in that habitat

Table 3: Examples of the habitats from which actinomycetes with antifungal and biological control activities were isolated

Habitat type	Antifungal activity against
Strawberry fields	*Botrytis cinerea*, Powdery mildew
Avocado mulch	*Phytophthora* spp.
Banana plantations	*Phytophthora* spp., *Phytium* spp., *Fusarium* spp
Macadamia orchards	*Phytophthora* spp.
Mango fields	*Phytophthora* spp.
Wheat fields	*Rhizoctonia* spp., *Trichoderma* spp.
Carrot fields	*Phytium* spp.
Woody scrubs	*Armillaria* spp.
Soils for organic farming	*Fusarium* spp., *Phytium* spp.
Vineyards	*Botrytis cinerea*, Powdery mildew
Tree barks	*Phoma* spp.
Allocasuarina rhizosphere soil	*Phytophthora* spp., *Armillaria* spp.
Cut flower-potting mixes	*Puccinia* spp., *Phytophthora* spp., *Trichoderma* spp.
Citrus orchards	*Phoma* spp., *Phytophthora* spp., *Collettotrichum* spp., *Alternaria* spp., *Penicillium* spp.

In addition to bioremediation purposes, the enzyme producing actinomycetes from the library such as keratinase producing *Nonomuraea* spp. isolated from poultry manure are also used to degrade chicken feathers and produce an environmentally friendly biofertilizer. Composting of pig and poultry manure is achieved using actinomycetes with recorded enzymatic activity.

4.1.4. Bacteriophages

Bacteriophages isolated towards the type strains of a wide range of actinomycete species as well as marine and freshwater bacteria in the Microbial Library are used

- as indicators to detect the rare actinomycetes in selected test substrates.
- as tools to determine the relatedness of the isolated taxa.

Table 4: Recoded enzymatic activities of the selected actinomycete taxa from the Sunshine Coast Region

Enzymes						
Genera	**Caseinase**	**Starch hydrolysis**	**Cellulase**	**Xylanase**	**Ligninase**	**Keratinase**
Streptomyces	+	+	+	+	+	+
Saccharomonospora	+	+	+	+	+	+
Actinomadura	+	+	+	+	+	+
Nonomuraea	+	+	+	+	+	+
Nocardioides	+	+	+	+	+	+
Nocardiopsis	+	+	+	+	+	+
Lechevalieria	+	+	+	+	+	+
Thermomonospora	+	+	+	+	+	+
Saccharothrix	+	+	+	+	+	-
Glycomyces	+	+	+	+	-	-
Streptosporangium	+	+	+	+	-	-

- ➢ to deselect unwanted taxa on isolation plates in the process of target specific selective isolations [3].
- ➢ to identify antiviral activity of the fermentation extracts of the isolates.

For example, detection and isolation of enzyme producing *Saccharomonospora* species were achieved using bacteriophages as indicators as well as aiding the selective isolation procedures [3]. Antiviral compound producing *Micromonospora* species were also isolated and tested for antiviral activity from the near shore marine sediments in the Sunshine Coast Region using the bacteriophages.

5. Contribution of the Microbial Library to the State's Biodiscovery Initiative

The Microbial Library at the University of the Sunshine Coast is thus expected

1. to contribute to the ongoing efforts of the Queensland Government to map Queensland's unique microbial diversity to attract biotechnological investment.

2. to complement the studies conducted by other research institutes in the State towards the establishment of a new "Microbial Resources Research Network" in Australia.
3. to provide a reservoir of strains for future work on the production of useful antibiotics, enzymes, and other biologically active compounds for potential commercialization in the State.
4. to support the development of state and nationwide policies for the conservation of the unique Australian microflora.

References

1. Biotechnology in Queensland, Australia. *Brilliant Science-Four billion years in the making. The Smart State Report 2001*. Queensland Government, Department of Innovation and Information Economy.
2. Kurtböke, D.I. (2000). Australian Actinomycetes; an unexhausted source for biotechnological applications. *Actinomycetologica*, **14**: 43-53.
3. Kurtböke, D.I. (Ed.) (2003). *Selective Isolation of Rare Actinomycetes*. Queensland Complete Printing Services, Australia.
4. Kurtböke, D.I. (2004). Uniqueness of the "Smart State's" Microbial Diversity, *From an Actinomycete Collection to Biodiscovery at the* University of the Sunshine Coast. *Microbiology Australia*, **25**: 36-38.
5. Sly, L. (1998). Australian Microbial Diversity. *Microbiology Australia*, **19**: 27-35.
6. Queensland Biodiscovery Bill, Fact Pack (2003). Queensland Government, Department of Innovation and Information Economy.

Actinomycetes taxonomy in the Aventis strain collection

J. M. Wink

Aventis Pharma Germany, Drug Innovation and Approval, 65926 Frankfurt am Main, Germany

1. Introduction

The **Aventis** microbial strain collection is located in Frankfurt am Main, Germany. The collection contains several thousand isolates of Actinomycetales stored in liquid nitrogen, deep freezers (-140°C to -40°C) and as freeze dried ampoules. The strains are isolates from highly diverse areas for natural product isolation. All new isolates were obtained under the Biodiversity Convention of Rio. The collection also includes about 90% of the validly described species of the Actinomycetales. The sections below describe the role of the taxonomy in this collection.

2. Control of microbial diversity in the Collection

The first step is to decide which taxonomic marker will be used for the characterization of such a huge number of Actinomycetes in the Aventis strain collection. A good taxonomic marker is characterized by the following points:

- Ubiquitary distribution within the Actinomycetales
- Large amount of information
- Reproducible analysis
- Simple and fast determination
- Fast identification result
- Low cost
- Automatised

Based on these characteristics, we chose the method of analysis of the fatty acids in cell wall as our taxonomic marker [3]. Therefore, we developed a rapid method based on the method of Müller *et al.* [5] and our own database.

The isolate is incubated in ISP 2 broth for 7 days at 28°C and 1 ml is transferred to the centre of a sterile Satorius filter (SM 11106) placed on an agar medium containing starch (10g/l), glucose (10g/l), glycerol (10g/l) cornsteep liquor (2,5 g/l), peptone (5g/l), yeast extract (2g/l), NaCl (1g/l) and $CaCO_3$ (3g/l). Incubation is carried out for 5 days at 28°C. Two loops of cell material are transferred into 10ml of distilled water in a microtube, 35 ml of methanolic TMSH is added and the sample is mixed. The sample is dryed with a nitrogen stream (Barkey Evaporator) and extracted into 100 ml of a mixture of 9 volumes of *tert*-butyl methyl ether and *n*-hexane (1:1, v/v) and one volume of methanol. The extract is directly used for the GC analysis, which is done with an HP 6890 GC (Hewlett-Packard).

With this method, it has so far been possible to determine about 20% of the *Aventis* strain collection up to the genus level. Thereby, we found that only 40 to 50% of the collection belongs to the genus *Streptomyces*. More than 30% belong to the *Micromonosporaceae*, *Streptosporangiaceae* and *Pseudonocardiaceae*. As a result, this collection, also growing by isolation of new strains over the years and by acquisition of other collections, is showing a high lever of diversity.

Beside this established method we also look for the implication of new molecular approaches for the dereplication of actinomycetes strains, for the establishment of correlations between genetic polymorphism and metabolic diversity and a fast typing and identifying of actinomycetes.

Therefore different approaches are tested:

- DNA extraction from actinomycetes strains, amplification and sequencing of the first 500 nucleotides of the 16S rDNA
- Analysis of the molecular polymorphisms generated after the digestion by restriction enzyme (REA) of the 16S rDNA, the heat shock protein 65 (*hsp65*) and the 16S-23S ribosomal intergenic spacer region (16S-23S ITS)
- The RiboPrinter®

From these three approaches the RiboPrinter is showing the best results for solving problems on the species level. The DNA sequencing of parts of the 16 S rDNA shows no good separation within different species of one genus, so we found homologies in these sequences higher than 90% within one genus such as *Actinomadura*. Also, the second approach did not lead to clear separations between the different isolates. Only the approach together with the German Culture Collection

(DSM) that used the RiboPrinter® for differentiation between isolates within one genus was very successful.

On the basis of the principle of riboprinting, an automated system has been developed, which is named RiboPrinter®. The RiboPrinter® system uses genetic "fingerprints" to identify and characterize microorganisms [1, 2, 8]. A simple colony pick is used to collect samples from agar plates. Each sample is suspended in buffer and heat treated in a sample carrier. After the carrier is transferred to the machine, all subsequent actions are fully automated. In the first step within the characterization unit, samples are treated with a lysing agent to release the DNA. The DNA is then digested to completion with a restriction enzyme.

The RiboPrinter® system transfers the resulting DNA restriction fragments to an agarose gel cassette containing thirteen wells. Samples are placed in eight of the wells; marker DNA of known molecular weights, included to establish a reference, occupies the remaining five wells. Using a process known as direct blot electrophoresis, the DNA fragments are separated by size and transferred to a moving nylon membrane. After denaturation, each membrane is hybridised with a chemically labelled rRNA operon from *Escherichia coli*. The membrane is washed and treated with blocking buffer and an antisulfonated DNA antibody/alkaline phosphatase conjugate. Unbound conjugate is removed and then a chemiluminiscent substrate applied. This step makes each electrophoresis band containing genetic information from the rRNA genes visible to a custom CCD camera in the RiboPrinter® system.

The camera detects the light intensity of the targeted bands and converts the patterns from luminescent DNA fragments to digital information. This image data is stored on the system computer's hard drive. The RiboPrinter® Microbial Characterization System analysis software uses the position and intensity of the five well characterized marker fragments, run alongside the samples, to normalize the resulting output data. The positions of this standard marker bands are used to correct for lane-to-lane and membrane-to-membrane variations in band position. Then the system generates the RiboPrint® pattern for each sample and marker lane using appropriate algorithms.

We studied 95 strains from our collection, which had been all classified as "*Actinomadura*" species based on chemo taxonomical markers like cell wall amino acids and sugars. All strains were treated with a method developed for *Streptomyces* strains and the DNA was digested with the enzyme PstI. 86 from the 95 strains show a RiboPrint pattern.

Strains with no or only a small number of bands in the pattern were investigated again by using the enzyme PvuII. Also with PvuII, only three strains show no pattern, all the others show patterns with more bands than by use of PstI for digest. Also by use of EcoRI as a third enzyme, two of the strains show no pattern in the RiboPrint. On the basis of the patterns after digest with PvuII, two small RiboGroups could be identified with strains of a similarity higher 0,94. In one group two strains and in the other three strains are included.

From 95 strains 93 show a RiboPrint pattern after using the two enzymes PvuII and PstI and from the 93 strains 90 show a different RiboPrint pattern. This indicated that there is a high diversity within the group of "*Actinomadura*" in the collection. A direct identification was not possible, because the RiboPrinter databases do not include the species of the genera *Actinocorallia, Actinomadura, Microteraspora* and *Nonomuraea* (former *Actinomadura)*.

3. New species = New metabolites

While screening for strains producing compounds active against methicilin resistant bacteria, a number of new species producing new antibiotics were isolated:

- *Amycolatopsis balhimycina,* glycopeptide antibiotic namened Balhimycin [13].
- *Amycolatopis keratiniphila* subsp. *nogabecina,* glycopeptide antibiotic named Nogabecin [13].
- *Amycolatopsis decaplanina,* glycopeptide antibiotic named Decaplanin [15]
- *Amycolatopsis vancoresmycina,* polyketide antibiotic named Vancoresmycin [13].
- *Actinoplanes friuliensis*, lipopetide antibiotic named Friulimicin [9]

Other novel species, which produce new compounds or are used for biotransformations, have been described in recent years:

- *Amycolatopsis tolypomycina,* producer of the ansamycin antibiotic Tolypomycin and used for biotransformations [13].
- *Nonomuraea ditzii,* biotransformations of polyene antibiotics like Nigericin [6]
- *Actinomadura namibiensis,* antiviral active peptide [12].

All these examples show that novel microorganisms often produce new metabolites. Also by modification of metabolites with biotransformation, novel microorganisms were found.

Fig.1: Comparison of different restriction enzymes by characterization of isolates of the genus *Actinomadura* with the RiboPrinter®

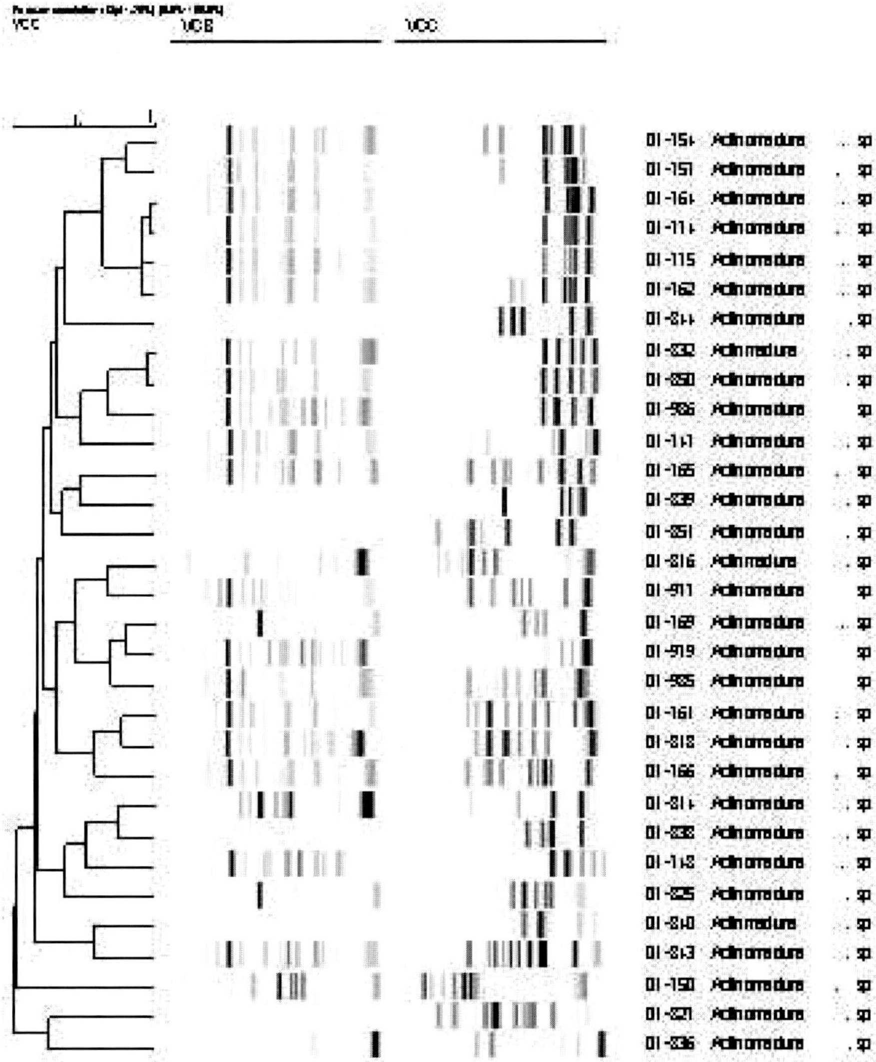

Figure 2: Sporangium of *Actinoplanes friuliensis* sp. Nov. (SEM, left) and the structure of Friulimicin

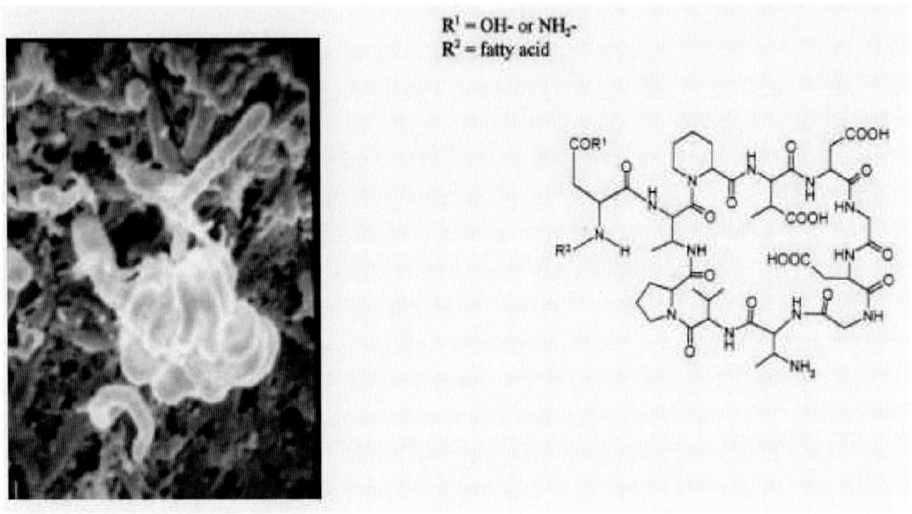

Resulting from the problematic species status in *Streptomyces* the novel isolates like the producer of the Cyclipostins, new hormone-sensitive lipase inhibitors, was only described as *Streptomyces* sp. [10].

In other cases novel species could not be described because the patent situation is still open and for the publication of a novel species the strain has to be deposited at two open collections.

Also the validation of species which have not been validly described and in which we found new compounds like ala(O)-actagardine [7], a new lantibiotic from cultures of *Actinoplanes "liguriae"*, is part of the taxonomic work [15].

4. Correlation between taxonomy and metabolite formation

The correlation of metabolite and antibiotic production within the members of the different genera of the *Pseudonocardiaceae* gives a much clearer picture as in the case of the members of the genus *Streptomyces*. Closely related species produce closely related antibiotics. In the genus *Amycolatopsis* we found one group producing glycopeptide antibiotics and one group producing ansamycin antibiotics.

Molecular biological analyses such as 16S rDNA sequencing (see fig. 3) and ribotyping show that two clusters could be defined [13].

The "mediterranei" cluster includes strains, which produce ansamycin antibiotics of the Rifamycin group like *Amycolatopsis vancoresmycina* (Homorifamycin), *Amycolatopsis tolipomycina* (Tolipomycin) and *Amycolatopsis mediterranei* (Rifamycine). Also *Amycolatopsis balhimycina,* a glycopeptide producer is placed in this group. In this strain the group of Prof. Wohlleben of the University of Tübingen, who works with the bisynthese genes of the glycopeptide Balhimycine, identified genes, which belong to the rifampicin biosynthesis cluster (personel communication).

The "orientalis" cluster embraces the glycopeptide producers like *Amycolatopsis alba, Amycolatopsis coloradensis, Amycolatopsis decaplanina, Amycolatopsis orientalis, Amycolatopsis lurida* and *Amycolatopsis keratiniphila* subsp. *nogabecina.* The strains and their metabolites are listed in Figure 4.

5. Exclusion of pathogenic actinomycetes

The exclusion of pathogenic Actinomycetes is a very important point for the personal safety of our co workers. The following diseaeses are caused by Actinomycetes [4]:

Figure 3: 16s rDNA dendrogramm of members of the genus *Amycolatopsis*

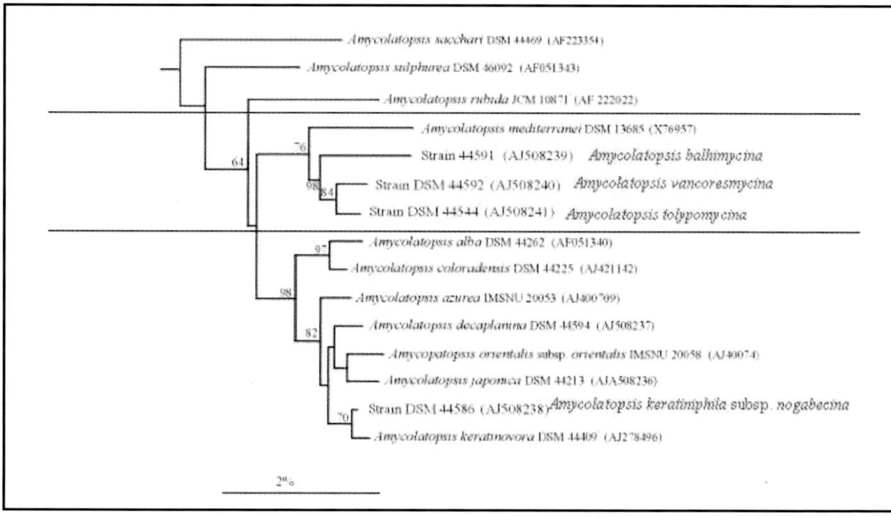

Figure 4: Metabolites described from the different *Amycolatopsis* species

Organism	Metabolite	Compound class
Amycolatopsis alba		Glycopeptide antibiotic
Amycolatopsis azurea	Octacosamicin	Antifungal Polyketide
Amycolatopsis balhimycina	Balhimycin	Glycopeptide antibiotic
Amycolatopsis coloradensis	Avoporacin	Glycopeptide antibiotic
Amycolatopsis decaplanina	Decaplanin	Glycopeptide antibiotic
Amycolatopsis japonica	Dethymicin	Immunosuppressant
Amycolatopsis keratiniphila subsp. ***nogabecina***	Nogabecin (Actinoidin B)	Glycopeptide antibiotic
Amycolatopsis mediterrenei	Rifamycin B,G,L,O,SV,W,Y	Ansamycin-type antibiotics
	Proansamycin B-M1	Ansamycin related antibiotic
	Homorifamycin	Ansamycin-type antibiotics
	Kanglemycin A	Ansamycin-type antibiotic
	Protorifamycin	Intermediate of Ansamycinbiosynthesis
Amycolatopsis orientalis subsp. *orientalis*	Vancomycin	Glycopeptide antibiotic
	Orienticin	Glycopeptide antibiotic
	Quartromicin	Macrocyclic antibiotic
Amycolatopsis orientalis subsp. *lurida*	Ristomycin	Glycopeptide antibiotic
	Benzanthrin	Quinone antibiotic
Amycolatopsis sulphureus	Chelocardin	Tetracycline antibiotic
	Rifamycin	Ansamycin-type antibiotic
Amycolatopsis tolypomycina	Tolypomycin	Ansamycin-type antibiotic
Amycolatopsis vancoresmycina	Homorifamycin	Ansamycin-type antibiotic
	Vancoresmycin	Plyketide-antibiotic

- **Nocardiosis**

➢ pulmonary nocardiosis

➢ systemic nocardiosis

➢ central nervous nocardiosis

➢ extrapulmonary, localized nocardiosis

➢ cutaneus, subcutaneus, and lymphocutaneus nocardiosis

➢ nocardial mycetoma

In temperate regions of the world the most frequently recognized form of disease is pulmonary nocardiosis caused by *N. asteroides*. In tropical and subtropical regions mycetoma caused by *N. brasiliensis.*

- **Actinomycetoma**

Chronical, tumorlike processes with involvement of the bones, formation of nodular lesions.

(Nocardia brasiliensis, Actinomadura madurae, Actinomadura pelletieri, Streptomyces somaliensis, Nocardia otitidisvaricarium, Nocardia transvalensis, Nocardia asteroides and *Nocardia farcinica)*

- **Dermatophilosis**

Dermatophilosis is an acute to chronic suppurative cutaneous disease that affects a wide range of animals as well as humans. Although dermatophilosis is recognized most frequently in livestock (i.e. cattle, sheep and horses) in tropical and subtropical areas of Africa, it has been diagnosed in most regions of the world. (*Dermatophilus congolensis)*

- **Allergic alveolitis**

Pneumonia with fever and shortness of breath, which later change to lung fibrosis. (*Saccharopolyspora rectivirgula, Saccharomonospora viridis* etc.)

Within the filamentous growing Actinomycetes we find only this small number of potential pathogenic species (L2–German Classification), which can be determined by their cell wall, physiology and morphology.

Also in the nonfilamentous growing genera of Actinomycetales we find a lot of pathogenic species in genera like *Arthrobacter, Brevibacterium, Cellulomonas, Corynebacterium, Dermabacter, Gordonia, Kocuria, Microbacterium, Mycobacterium, Rhodococcus, Tsukamurella, Turicella* etc.

For the determination of this species, the Biolog® system is a very helpful tool.

The Biolog GP2 MicroPlate (Figure1) is designed for identification and characterization of a very wide range of aerobic gram-positive bacteria. Biolog´s MicroPlates and databases were first introduced in 1989, employing a novel, patented redox chemistry. This chemistry, based on the reduction of tetrazolium, responds to the process of metabolism (i.e. respiration) rather than to metabolic by-products (e.g. acid). Biologs´s chemistry works as a universal reporter of metabolism and simplifies the testing process given that color developing chemicals do not need to be added. Since the GP2 MicroPlate is not dependent upon growth to produce identifications, it provides superior capability for all types of gram positive organisms: cocci, rods, and spore-forming rods all are identified with a single panel. The database for the GP2 MicroPlate is now over 310 species. It is by far the largest kit-based identification database available.

The Biolog GP2 MicroPlate performs 95 discrete tests simultaneously and gives a characteristic reaction pattern called a “metabolic fingerprint”. These fingerprint reaction patterns provide a vast amount of information conveniently contained in a single Biolog MicroPlate. The metabolic fingerprint patterns are compared and identified using the MicroLog™ database software.

Other aerobic kit-based identification methods rely on much smaller number of tests. Consequently, the significant limitation of these products is the limited number of species and organism types that they can identify. Furthermore, these products were designed to address the needs of routine clinical/hospital testing. The Biolog GP2 MicroPlate was designed to address the needs of much wider range of users including environmental testing labs and animal and plant disease labs as well as clinical reference labs.

Procedure for using GP2 MicroPlates:

Figure 5: Biolog pattern of different Actinomycetes

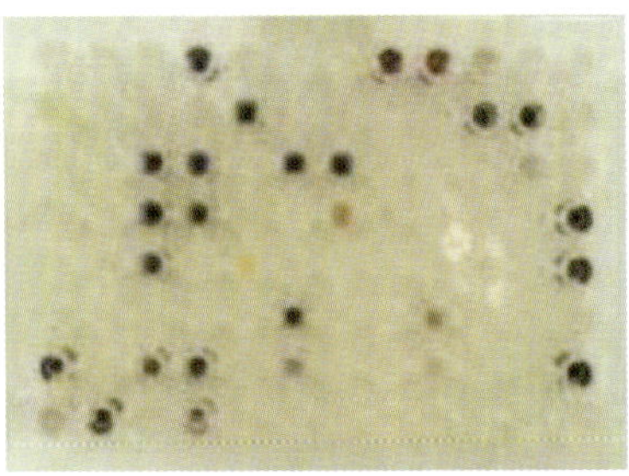

Arthrobacter woluwensis

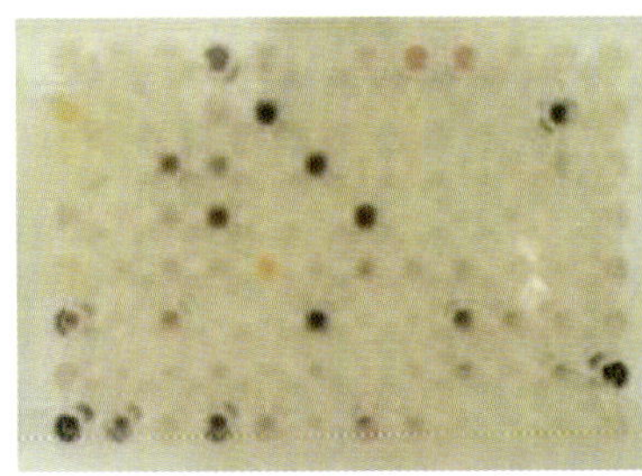

Arthrobacter luteolus

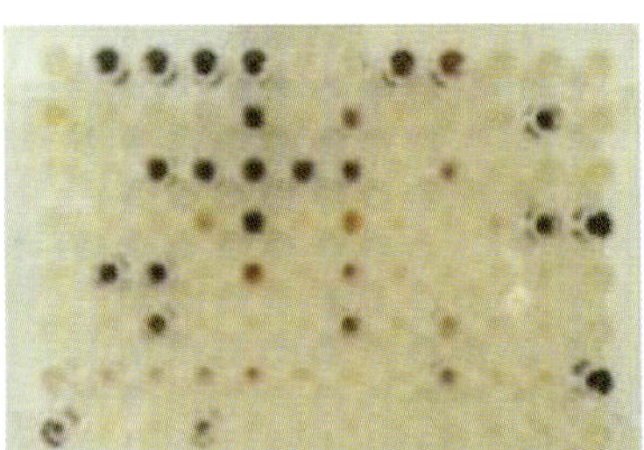

Microbacterium chocolatum

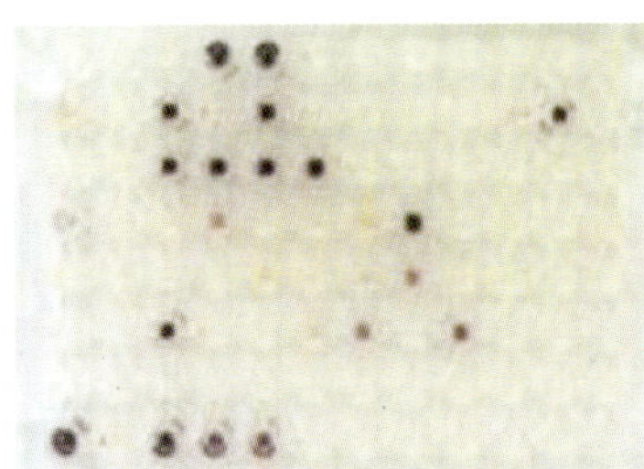

Microbacterium halophilum

A pure culture of a bacterium is grown on a Biolog Universal Growth agar plate (BUG-Agar). The bacteria are swabbed from the surface of the agar plate, and suspended to a specified density in GP inoculating fluid 150 ml of bacterial suspension is pipetted into each well of the GP2 MicroPlate.

The MicroPlate is incubated at 30° or 35°C (depending upon the nature of the organism) for 4-24 hours. Finally, the MicroPlates are read either visually or with the Biolog MicroStationäSystem and compared with the GP database.

References

1. Bruce, J. L. (1996). Automated system rapidly identifies and characterizes microorganisms in food. *Food Technology*, **50** (1): 77-81.

2. Bruce, J. L., Hubner, R.J., Cole, E.M., McDowell, C.I. and Webster, J.A. (1995). Sets of EcoRI fragments containing ribosomal RNA sequences are conserved among different strains of *Listeria moncytogenes*. *Proceedings of the National Academy of Science USA*, **92**: 5229-5233.

3. Kroppenstedt, R.M. (1985). Fatty acid and menaquinone analysis of Actinomycetes and related organisms. In *Chemical Methods in Bacterial Systematics* (Goodfellow, M. and Minnikin, D.E., Eds.). No. 20 SAB Technical Series, Academic press. pp.173-199.

4. Lennette E.H., Balows, A., Hausler, W.J. and Shadomy, H.J. (1985). *Manual of Clinical Microbiology*. 4th edition. American Society of Microbiology Washington D.C.

5. Müller, K.D., H. Husmann and H.P. Nalik. (1990). A new and rapid method for the assay of bacterial fatty acids using hight resolution capillary gas chromatography and trimethylsulfonium hydroxyde. *Zentralblatt Bakteriologie*, **274**: 174-182.

6. Stackebrandt, E., J. Wink, U. Steiner, and R.M. Kroppenstedt. (2001). *Nonomuraea dietzii* sp. nov. *International Journal of Systematic and Evolutionary Microbiology*, **51**: 1437-1441.

7. Vertesy, L., W. Aretz, A. Bonnefoy, E. Ehlers, M. Kurz, A. Markus, M. Schiell, M. Vogel, J. Wink and H. Kogler. (1999). Ala(O)-actagardine, a New Lantibiotic from Cultures of *Actinoplanes liguriae* ATCC 31048. *Journal of Antibiotics*, **52**: 730-741.

8. Webster, J. A., Bannerman, T.L., Hubner, R. J., Ballard, D. N., Cole, E. M., Bruce, J. L., Fiedler, F., Schubert, K., and Kloos, W. E. (1994). Identification of the *Staphylococcus sciuri* Species Group with EcoRI Fragments Containing rRNA Sequences and Description of *Staphylococcus vitulus* sp. nov. *International Journal of Systematic Bacteriology*, **44**: 3: 454-460.

9. Wink, J., Aretz, W., Meiwes, J., Seibert, G. and Vobis, G. (2000). Friulimicins: Novel lipopeptide antibiotics with peptidoglycan synthesis inhibiting activity from *Actinoplanes friuliensis* sp. nov., taxonomic studies of the producing microorganism and fermentation. *Journal of Antibiotics*, **53**: 807-815.

10. Wink, J., Schmidt, F.-R., Seibert, G. and Aretz, W. (2002). Cyclipostins: Novel Hormone-Sensitive Lipase Inhibitors from *Streptomyces* sp. DSM 13381. Taxonomic studies of the producer microrganism and fermentation results. *Journal of Antibiotics*, **55**: 472-479.

11. Wink, J., Kroppenstedt, R.M., Seibert, G. and Stackebrandt, E. (2003). *Actinomadura namibiensis* sp. nov. *International Journal of Systematic and Evolutionary Microbiology*, **53**: 721-724.

12. Wink, J., Kroppenstedt, R.M., Ganguli, B.N., Nadkarni, S.R., Schumann, P, Seibert, G. and Stackebrandt, E. (2003). Three new antibiotic producing species of the genus *Amycolatopsis, Amycolatopsis balhimycina* sp. nov., *A. tolypomycina* sp. nov., *A. vancoresmycina* sp. nov. and description of *Amycolatopsis keratiniphila* subsp. *keratiniphila* subsp. nov. and *A. keratiniphila* subsp. *nogabecina* subsp. nov. *Systematic and Applied Microbiology*, **26**: 38-46.

13. Wink, J. (2003). Polyphasic taxonomy and antibiotic formation in some closely related genera of the family *Pseudonocardiaceae.* In *Recent Research Developments in Antibiotics* (Pandalai, S.G., Ed.). Transworld Research Network. Kerala India.

14. Wink, J., Gandhi, J., Kroppenstedt, R.M., Seibert, G., Straubler, B., Schumann, P. and Stackebrandt, E. (2004). *Amycolatopsis decaplanina* sp. nov., a novel member of the genus with unusual morphology. *International Journal of Systematic and Evolutionary Microbiology,* **54**: 235-239.

15. Wink, J., Kroppenstedt, R.M., Schumann, P., Seibert, G. and Stackebrandt, E. (2004). *Actinoplanes liguriae* sp. nov., and *Actinoplanes teichomyceticus* sp. nov., two new species of the genus *Actinoplanes* (in preparation).

Industrially-held microbial diversity: Are culture collections really necessary?

D. D. Baker

Albany Molecular Research, Inc., Bothell Research Center, 18804 North Creek Parkway, Bothell, Washington 98011-8012 USA

1. Introduction

Natural product discovery is the process of identifying chemical compounds created by biological organisms as useful therapeutic agents for overcoming human, animal or plant diseases. A large percentage of ethical (prescription) drugs on the world market are themselves natural products or are derived from natural products [2,3]. Although rooted in ethnomedicinal traditions, modern natural product discovery did not become mainstream science until the early 20th century, and then dominated the field of pharmaceutical research through the latter part of that century. Programs in molecular or genetic engineering, rational drug design, combinatorial chemistry, etc. have been proposed to replace natural product discovery, but to date none of these fields has contributed so much to the development of medicines as natural product discovery.

1.1 Trends in the pharmaceutical industry

Despite the large contribution natural product discovery has played over the past century, major pharmaceutical companies have been eliminating programs in natural product discovery during the past few years and also eliminating the associated microbial culture collections thereof (see Table 1).

These industrial collections, which have taken decades to build and characterize, are seen to have little or no value. The cost of maintaining such collections, with little hope to use them in the future, is seen to be prohibitive. As a result, fairly significant collections of actinomycetes, filamentous fungi, myxobacteria and some cyanobacteria are now threatened with extinction. In reality, do such collections really need to be protected? Most likely, all of the biodiversity present in these threatened collections is accessible from nature through additional isolations. Also, if one takes a genetic approach, the inherent diversity of genes represented in

Table 1: Major pharmaceutical companies and natural product discovery (NPD) programs

Companies eliminating internal NPD	Companies continuing some internal NPD
Glaxo-SmithKline *	Wyeth
Abbott *	Novartis
Hoffman-La Roche *	Aventis
Eli Lilly & Company *	Merck & Company
Bristol Myers Squibb	
Pfizer (including Pharmacia, but excluding animal health)	

* These companies have disposed of their microbial collections

threatened microbial collections is undoubtedly readily available from other publicly-supported strain collections, or from genomic DNA isolated directly from environmental samples. Therefore, the potential loss of industrial culture collections might not be considered a dire predicament which needs to be resolved. Instead, it might be argued that the funds required to maintain culture collections could be diverted to investments in genetic, proteomic or virtual technologies wherein the majority of therapeutic agents could be predicted to arise.

If one follows recent trends in the pharmaceutical or agricultural industries, it would appear that natural product discovery will not play a major role in drug discovery in the foreseeable future. Natural product samples are considered too complex and logistically difficult to work with. Synthetic compounds are much simpler to work on, do not have the structural rigidity that natural product scaffolds bear, and can quickly be manipulated using traditional medicinal chemistry approaches. There is also the belief that natural products are limited in their utility to cytotoxic agents, useful for antibiotics and anticancer drugs, but not useful for other disease indications. The combined revenue value of natural product drugs has continued to drop for the past decade or so. Genomic and proteomic studies are anticipated to identify the majority of disease targets in the human body within the next few years. Then, computer-aided drug design (CADD) programs are expected to identify novel chemotypes for therapeutics from computer modeling of molecule-target interactions in the absence of large screening libraries. Perhaps it is now time to recognize that microbial strain collections are antiquated hobbies for intellectual microbiologists. Modern "biological" collections can be stored in electronic form where select product pathways can be chosen for cloning and production.

The shortfall in this reductionist view is that biological systems, whether derived from nature or created in the laboratory, are inherently unpredictable. Scientists cannot yet design drugs for each and every disease known to afflict humans, animals or plants. Known diseases, like tuberculosis or malaria, evolve over time to become resistant to those drugs (e.g., rifampin and quinine, respectively) which initially controlled them. If scientists had been successful in planning for and overcoming resistance, diseases like tuberculosis would not have become the pandemics that they currently are. Emerging diseases, like SARS, BSE or haemorrhagic fevers, will pose new and unknown threats that might not be able to be addressed solely from a "design approach" perspective. For most design approaches, having a bioactive molecule is essential or at least helpful in jump-starting the program.

Discovery often has an element of serendipity. Biological systems, being so very complex and evolved to interact with other biological systems, will constantly give rise to novel chemical structures that will be selected naturally to be effectors of molecular interactions. By continuing to screen extracts from natural product samples, novel drug candidates can and will be found. Newman *et al.* [3] clearly demonstrate that natural product derived drugs have utility in numerous therapeutic indications (see Figure 1), not just as antibiotics. Therefore, is it more efficient to isolate new strains each time assay screening is desired, or should culture collections be the source of natural product extracts?

1.2 Implications for the preservation of microbial culture collections

Albany Molecular Research has estimated that the cost of maintaining existing isolated microbial strains under liquid nitrogen vapor conditions is approximately US$ 0.15 per strain per year, for a collection of 150,000 strains, including the personnel costs for curation of the collection. If redundancy of the collection (storage at a second secure site) is factored in, the cost rises to US$0.30 per strain per year for this size of collection. Conversely, we also have estimated that the cost to isolate new strains from Rio-compliant environmental samples will be on average about US$30.00 per strain. Therefore, one can rationalize storage of existing collections of strains for numerous decades before the cost of storage exceeds the cost of reisolating strains for natural product library development (see Table 2). By

Figure 1. Natural product new drugs by indication for the years 1981-2002. (Data adapted from Newman *et al.* [3])

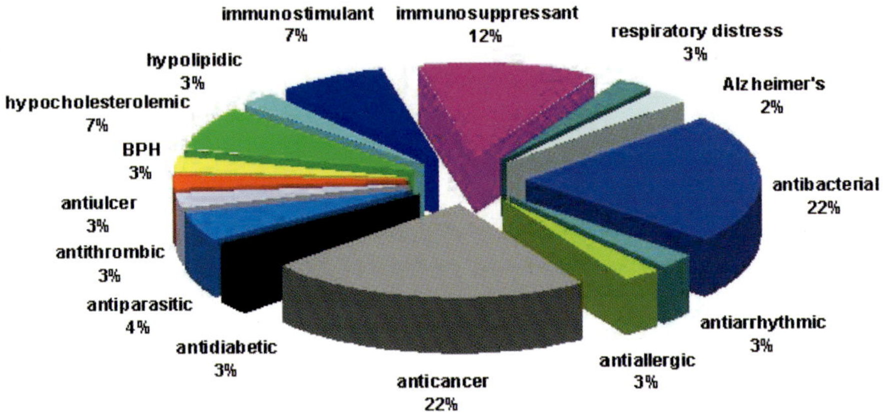

Table 2: Maintenance versus acquisition costs for industrial microbial collections

	Maintenance of existing collection	**Isolation of new cultures to create a collection**
Isolation source materials, reagents, labor, etc.		US$30/strain
Liquid nitrogen freezers, LN_2, curation, etc.	US$ 0.30/strain	
Number of strains per year	X 150,000	X 2,000
Number of years	X 20	X 20
Number of strains at conclusion of period	150,000	40,000
Total costs	US$ 900,000	US$ 1,200,000
Cost per year	US$ 45,000	US$ 60,000

extrapolation, if a microbial collection gives rise to a commercialized drug once every decade, and the revenue from the drug is only modest, the collection will have paid for itself.

Even if the pharmaceutical industry determines that continuing to hold existing collections is not warranted for business purposes, it must recognize that publicly held collections cannot serve a discovery role if they return to natural product discovery programs in the future. Publicly held collections, by the very nature of their founding and support, may not have sufficient strain diversity or depth to support natural product discovery. The majority of pharmacologically relevant molecules have been derived from the filamentous fungi, the actinomycetes, a few eubacterial groups and to a much lesser extent, cyanobacteria and myxobacteria. Public culture collections are rich in selected microbial groups, and are particularly valuable as repositories of taxonomically characterized microorganisms, but they may not have sufficient phylogenetic depth to support natural product discovery programs or build fermentation extract libraries. Although taxonomic characterization and diversity are important to public culture collections, a far more important factor for industrial collections is to have a selection of strains which provide both breadth and depth of chemical diversity.

2. Technological improvements making natural product discovery more efficient

New analytical technologies are also making discovery of microbial-derived metabolites simpler and quicker. Improvements in chromatography, mass spectrometry and robotic instrumentation can strengthen and expand the capabilities of the natural product chemist. Identification of chemical structures of microbial metabolites can be achieved in hours and days, rather than the weeks and months required a decade ago (see Figure 2). However, academic chemistry departments have reported a decrease in programs focused on natural product discovery research. Just as the technology becomes available to make natural product chemistry more efficient, the limited pool of experienced scientists complicates the problem. The training of natural product chemists to utilize these newer technologies needs to be encouraged.

The novelty of natural product compounds is not universal. In conducting screening of natural product samples through biochemical or cell based assays, the result can be the identification of bioactive known compounds, i.e., compounds that have been previously published in the scientific literature. *Known* compounds can be less interesting to industry because they rarely can be the subject of composition of matter patents. However, this does not preclude them from being patented for use. Compounds identified through screening which have not been published in the scientific literature previously can be of two types. *New* compounds are those which are analogues or derivatives of chemical scaffolds that have already been published. Functional groups or the structural conformation may be different. Truly *novel* compounds are those for which the basic chemical scaffold is unique, and not represented in the scientific literature previously. Natural product screening also gives rise to *nuisance* compounds. These are chemicals which are identified as active in numerous types of assays, and because of this apparent non-specific activity, do not constitute good therapeutic lead compounds. An effective natural product discovery program will have protocols in place to quickly identify these different types of bioactive molecules, and prioritize those which have the greatest potential for characterization and development. This process is typically called de-replication, or sometimes de-convolution.

Another point which must be clearly recognized is that through the screening of natural product samples, it is more likely that a lead candidate molecule will be discovered, rather than a drug. For example, the drug Demerol (meperidine) is a molecule based on the plant-derived tropane alkaloid, atropine. During the very early days of natural product discovery, compounds which were identified to be bioactive against a particular disease became drugs administered to humans primarily without chemical modification. Examples are natural penicillin, streptomycin, gentamicin or actinomycin D. Although several of these molecules have significant side effects, the diseases for which they were used were so difficult to treat that any bioactive compound was better than nothing. Modern pre-approval testing of drugs for human use frequently requires significant testing and often chemical modification to prevent side effects, to improve availability or to increase specificity for the target disease [1]. For example, the drug Vepesid (etoposide) is a semi-synthetic derivative of podophyllotoxin, a toxic plant metabolite. This means that medicinal chemists will need to work with the natural product scaffolds that are discovered. This can be difficult because of the frequent occurrence of complexity in natural product molecules or especially difficult stereochemistry. Fortunately, the technologies for chemically optimizing natural product lead compounds have improved. If the natural product molecule is itself not suitable as a drug, there are several ways to study analogues or derivatives as optimized leads. A relatively simple approach, taking

Figure 2: Timeline for de-replication of bioactive hits arising from high throughput screening of natural product libraries.

Approach	Scale-up fermen-tation	Crude fraction-ation	Re-assay	Fine fraction-ation	Re-assay	Structure elucidation
Historical Methods	2-4 weeks	1 week	1-2 weeks	1 week	1-2 weeks	1-3 weeks (depending on complexity)
LC/MS method or pre-fractionated library	N/A	N/A	N/A	1 week	1-2 weeks	1-3 weeks (depending on complexity)
Affinity binding	N/A	N/A	N/A	combined process: 1-2 weeks		1-3 weeks (depending on complexity)

advantage of evolutionary chemical diversity, is to identify naturally-occurring analogues. This can be approached by looking at phylogenetically related microorganisms (or plants) to determine whether families of similar compounds might be isolated and tested for improved utility. A second method, again which takes advantage of natural chemical pathways, is to review microbial enzymes or whole cell systems which can modify the lead compound in ways not possible by synthetic chemistry routes. Such procedures are called biotransformation or biocatalysis, and significant progress has been made in this field of research. A third method for optimizing a natural product lead is to employ computer-aided drug design (CADD) methodologies. Computer assisted modeling and similarity studies can now look for pharmacophore and structural features of natural product compounds and identify mimetics that may have similar or improved bioactivities. Although the natural product molecule itself may not become a drug, the knowledge inherent in biologically-evolved systems can be instructive to develop alternate, and often synthetic, compounds with improved efficacy.

3. Conclusions

The rationalization of a microbial collection, and/or a natural product discovery program, cannot be made solely through an accounting or time management process. Prioritization of new programs, with the allure of novel therapeutic opportunities, may be chosen over the more traditional programs. Also market forces for more research and development may influence large pharmaceutical company decisions about maintaining programs internally. Success will be the ultimate test of natural product discovery programs over the next decade. Currently, they continue to provide good value. In 2001, 2002 and 2003 in the US alone, a single class of natural product derived drugs (statins) provided the greatest revenues of all prescription medicines, in excess of US$10 billion each year (see Figure 3). New important natural product drugs, like daptomycin, camptothecin, bryostatin, discodermolide or exendins, are continuing to filter through the clinic and be approved for important diseases. Small, committed groups of natural product scientists around the world are making interesting discoveries that will lead to novel medicines within the next few decades.

Figure 3: Revenues for sales of the top 20 prescription drugs in the U. S. Market.

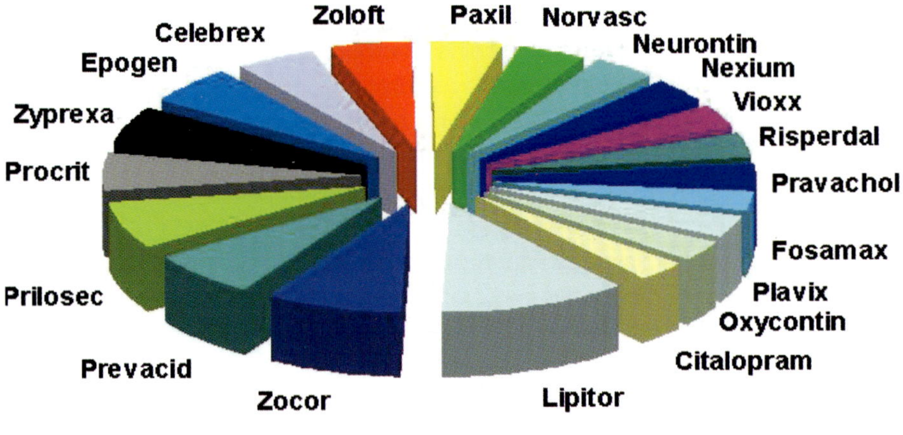

Lipitor, Zocor and Pravachol are natural products of the "statin" chemical family for cholesterol regulation. Oxycontin is an analogue of morphine (a plant-derived molecule) for pain. The remainders are synthetically derived chemicals.

Table 3: Selected providers of NPD services

Company	Microbial collections acquired (if any)
Albany Molecular Research, Inc.	Panlabs, Sterling-Winthrop, Lilly, Abbott
Analyticon Discovery	
Basilea Pharmaceuticals	Hoffman La Roche
BioLeads	
Cerylid Biosciences	
Drug Discovery, Ltd.	
MerLion Pharmaceuticals	Glaxo-Wellcome

Most of these groups are part of pharmaceutical industry service providers (see Table 3), and can be more flexible in addressing the special needs of natural product discovery programs than large pharmaceutical companies. One of the major reasons that these committed groups have been able to succeed is that they firmly believe in the value of natural products in medicine. They have understood that microorganisms, or plant or animal cells, as nature's teams of chemical synthetic factories, are more capable and diverse than any and all human chemists combined. The future for these programs looks bright.

Acknowledgments

The author thanks Khisal Alvi, Cheryl Garr and Ian Scott for carefully reviewing the draft manuscript and providing suggestions for improvements.

References

1. Jorgensen, W. L. (2004). The many roles of computation in drug discovery. *Science,* **303**: 1813-1818.
2. Newman, D. J., Cragg, G.M. and Snader, K.M. (2000). The influence of natural products upon drug discovery. *Natural Product Reports,* **17**: 215-34.
3. Newman, D. J., Cragg, G.M. and Snader, K.M. (2003). Natural products as sources of new drugs over the period 1981-2002. *Journal of Natural Products,* **66**: 1022-37.